Advances in Science, Technology & Innovation

IEREK Interdisciplinary Series for Sustainable Development

Advances in Science, Technology & Innovation (ASTI) is a series of peer-reviewed books based on important emerging research that redefines the current disciplinary boundaries in science, technology and innovation (STI) in order to develop integrated concepts for sustainable development. It not only discusses the progress made towards securing more resources, allocating smarter solutions, and rebalancing the relationship between nature and people, but also provides in-depth insights from comprehensive research that addresses the **17 sustainable development goals (SDGs)** as set out by the UN for 2030.

The series draws on the best research papers from various IEREK and other international conferences to promote the creation and development of viable solutions for a **sustainable future and a positive societal** transformation with the help of integrated and innovative science-based approaches. Including interdisciplinary contributions, it presents innovative approaches and highlights how they can best support both economic and sustainable development, through better use of data, more effective institutions, and global, local and individual action, for the welfare of all societies.

The series particularly features conceptual and empirical contributions from various interrelated fields of science, technology and innovation, with an emphasis on digital transformation, that focus on providing practical solutions to **ensure food, water and energy security to achieve the SDGs.** It also presents new case studies offering concrete examples of how to resolve sustainable urbanization and environmental issues in different regions of the world.

The series is intended for professionals in research and teaching, consultancies and industry, and government and international organizations. Published in collaboration with IEREK, the Springer ASTI series will acquaint readers with essential new studies in STI for sustainable development.

ASTI series has now been accepted for Scopus (September 2020). All content published in this series will start appearing on the Scopus site in early 2021.

Khanneh Wadinga Fomba · Bertrand Tchanche Fankam · Abdelwahid Mellouki · Daniel M. Westervelt · Michael R. Giordano
Editors

Advances in Air Quality Research in Africa

Proceedings of the First International Conference on Air Quality in Africa (ICAQ'Africa 2022)

Springer

Editors
Khanneh Wadinga Fomba
Department of Atmospheric Chemistry
Leibniz Institute for Tropospheric Research
(TROPOS)
Leipzig, Germany

Abdelwahid Mellouki
Mohammed VI Polytechnic University (UM6P)
Ben Guerir, Morocco

Michael R. Giordano
Mechanical Engineering
Carnegie Mellon University
Pittsburgh, PA, USA

Bertrand Tchanche Fankam
Department of Physics
Alioune Diop University
Bambey, Senegal

Daniel M. Westervelt
Department of Physics
Lamont-Doherty Earth Observatory
of Columbia University
New York, NY, USA

ISSN 2522-8714 ISSN 2522-8722 (electronic)
Advances in Science, Technology & Innovation
IEREK Interdisciplinary Series for Sustainable Development
ISBN 978-3-031-53524-6 ISBN 978-3-031-53525-3 (eBook)
https://doi.org/10.1007/978-3-031-53525-3

This Springer imprint is published by the registered company Springer Nature Switzerland AG
The registered company address is: Gewerbestrasse 11, 6330 Cham, Switzerland

Paper in this product is recyclable.

Contents

Sources and Characterization of Air Pollutants

Assessment of Carbon Dioxide Emission Factors from Electricity Generation in Burkina Faso

Bernard Nana, Hamadi Zalle, Issoufou Ouarma, Tizane Daho, Arsène Yonli and Antoine Béré

Abstract

Electricity generation is the second largest emitter of greenhouse gases, notably carbon dioxide (CO_2), in the Burkinabe energy sector. Until now, Burkina Faso has used the default emission factors provided by the Intergovernmental Panel on Climate Change to assess its CO_2 emissions from electricity generation, particularly in the context of the National Communications on Climate Change, due to the lack of country-specific emission factors. This study presents the assessment of CO_2 emission factors of electricity generation in Burkina Faso. The Burkinabe National Electricity Company was chosen as the scope of the study. The evaluation of emission factors for combustion is based on the experimental/calculation method through an analysis of the fuels consumed by the thermal power plants. After evaluating a CO_2 emission factor of 76,903.31 kg/TJ for the combustion of Heavy Fuel Oil (HFO) and 73,524.54 kg/TJ for the combustion of Distillate Diesel Oil (DDO), we evaluate the CO_2 emissions of electricity production at 579.83 Gg. The CO_2 emission factor for thermal generation is estimated at 0.663 kg/kW h and that of the electricity generation mix is 0.569 kg/kW h. The use of renewable energies in electricity generation avoided 16.7% of CO_2 emissions in 2018. Also, there is a decrease in the emission factor of the generation mix with the increase in the share of renewables in the generation mix.

Keywords

Electricity generation · Greenhouse gases · Emission factor · Carbon dioxide

1 Introduction

Combating air pollution and the adverse effects of climate change is one of the most important environmental issues of our time. According to the Intergovernmental Panel on Climate Change (IPCC), the increase in the concentration of GHGs (carbon dioxide (CO_2), methane (CH_4), nitrous oxide (N_2O) etc.) in the atmosphere as a result of human activities is the main cause of global warming (GIEC, 2021). According to the 6th IPCC report, net global anthropogenic GHG emissions are estimated to be 59 $GtCO_2$ in 2019 (of which 44 Gt CO_2), i.e. an increase of about 54% compared to the 1990 level (GIEC, 2022).

GHG emissions from anthropogenic sources come from several sectors of activity, including the energy sector. In the energy sector, GHG emissions come mainly from the use of primary fossil fuels for electricity generation and transport. According to a report by the International Energy Agency (IEA), global primary energy consumption is estimated at 14.3 Gtoe in 2018, of which 80.55% is fossil-based; and energy-related CO_2 emissions in the same period reached 33.14 Gt, an increase of 1.7% compared to 2017, the majority of which comes from electricity generation (IEA, 2019a).

Africa's contribution to energy-related CO_2 emissions is about 2% of cumulative global emissions, yet it is at the foreground of climate change impacts (IEA, 2019b). West Africa is already experiencing loss of life, human health impacts, biodiversity loss, reduced economic growth etc. (CDKN, 2022).

B. Nana (✉) · H. Zalle · I. Ouarma · T. Daho · A. Yonli · A. Béré
Laboratoire de Physique Et de Chimie de L'Environnement, Université Joseph KI-ZERBO, Ouagadougou, Burkina Faso
e-mail: bernard.nana@ujkz.bf

B. Nana
Ecole Normale Supérieure, Koudougou, Burkina Faso

I. Ouarma
Centre Universitaire de Banfora, Université Nazi BONI, Bobo-Dioulasso, Burkina Faso

K. W. Fomba et al. (eds.), *Advances in Air Quality Research in Africa*,
Advances in Science, Technology & Innovation, https://doi.org/10.1007/978-3-031-53525-3_1

In the face of climate change, reducing CO_2 emissions is a major priority. The member states of the United Nations Framework Convention on Climate Change (UNFCCC) have committed to reducing their GHG emissions. According to the Paris Climate Agreement, the objective of this reduction is to keep the increase in global average temperature well below 2 °C above pre-industrial levels and to continue efforts to limit this increase to 1.5 °C (Nations Unies, 2015).

An important step in the search for solutions to reduce emissions is to identify the most significant sources of GHG emissions. This is done through an inventory of GHG emissions and the development of programs to reduce the consumption of petroleum products and to increase the consumption of low GHG emitting products.

Burkina Faso, a member of the UNFCCC and a country not included in Annex I of the Convention, has joined the GHG emissions reduction process. Indeed, it has prepared National Communications under the UNFCCC. These communications show that in 2015 the energy sector was the second largest GHG emitting sector and electricity production was the second largest contributor to GHG emissions in this sector after transport in 2015 (PRBA, 2021). CO_2 typically accounts for 95% of energy sector emissions (GIEC, 2006a). CO_2 alone accounted for more than 99% of GHG emissions from electricity generation in Burkina Faso in 2015 (PRBA, 2021).

Several methods for estimating GHG emissions exist in the literature (CITEPA, 2020): the continuous measurement method, the material balance method and the calculation method using emission factors. The IPCC provides three tiers for estimating GHG emissions from stationary combustion (GIEC, 2006b): Tier 1 is the basic method, Tier 2 is the intermediate method and Tier 3 is the most demanding method in terms of complexity and data requirements. Emission factors are specific to the emission source and are determined either by analogy (default value), life cycle analysis, or experimentally and by calculation (studies/research). CO_2 emission factors for fossil fuel combustion depend mainly on the carbon content of the fuel and its calorific value. The CO_2 emission factor may include the oxidation factor (case of national emission factors).

Until now, Burkina Faso has assessed its GHG emissions attributable to electricity production with the default EFs provided by the IPCC due to the lack of country-specific EFs. However, a rigorous estimation of the sector's contribution to GHG emissions requires quality EFs. It is therefore important to develop specific EFs for Burkina Faso in the power generation sector. Moreover, as the reduction of GHG emissions is one of the key areas for combating climate change, the challenge for all countries is to implement a transition to a safer and less CO_2 emitting energy system such as renewable energy without hampering socio-economic development. It is in this context that the objective of this study is to assess the CO_2 emission factors of the electricity generation sector in Burkina Faso.

2 Materials and Methods

2.1 Burkinabe Electricity Generation

Burkina Faso's national electricity production in 2018 was 1052.8 GW h, of which 1020.7 GW h was produced by the Burkinabe National Electricity Company (SONABEL) (ME, 2018a). SONABEL's production is made up of 85.75% diesel thermal source; 8.95% hydro and 5.3% solar PV. The electrical energy delivered to distribution is 1,844.2 GW h and is provided by SONABEL (1,815.4 GW h) for the national interconnected network on the one hand and by the electricity cooperatives (28.8 GW h) for the isolated networks on the other (ME, 2018a, b). The study area considered is SONABEL because it is the main supplier of electricity in Burkina Faso with 97% of total national generation and nearly 96.5% of national thermal generation.

The number of functional power plants by type in 2018 was 19 for SONABEL (including 13 diesel thermal plants, 04 hydroelectric plants and 02 solar PV plants).

Figure 1 shows the location of SONABEL's power plants for the year 2018.

In 2018, thermal power plants consumed 186,805 tonnes of fuel (SONABEL, 2018). The share of HFO is 86.5% and that of DDO is 13.5%; that is, 161,586,325 kg of HFO and 25,218,675 kg of DDO.

2.2 Methodology

The CO_2 EFs of the fuels for combustion and the CO_2 EF of the electricity production of fossil thermal power plants depend on the characteristics of the fuels (carbon content, calorific value etc.) and do not depend on the technology used. The GHG emitting activities in this study are the activities of fuel combustion for electricity generation by power plants. The activity data used in this study are the actual quantities of fuels consumed and the electricity production for the year 2018.

3 Fuel Analysis by Characterisation

Molecular characterisation of fuels. A sample of the two fuels consumed by SONABEL's thermal power plants is taken and subjected to analysis. This characterisation provides the different constituents in terms of HFO and DDO

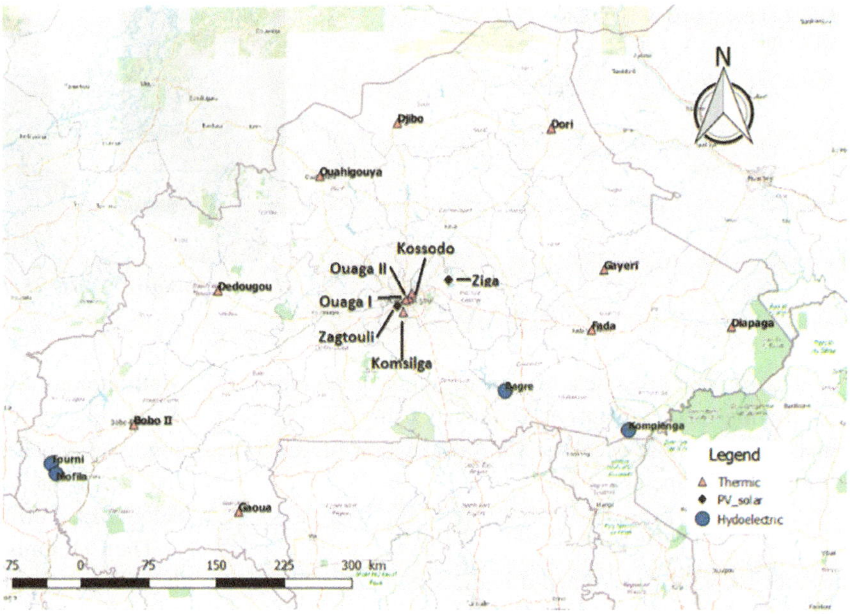

Fig. 1 Location of SONABEL's power plants in the country in 2018

molecules. The knowledge of the molecules constituting each fuel and their fraction in mass makes it possible to determine the chemical formula and the elementary composition in mass of the fuels. If $C_xH_yO_zN_uS_t$ is the chemical formula of the fuel, then:

$$x = \sum_{i=1}^{n} p_i x_i;\ y = \sum_{i=1}^{n} p_i y_i;\ z = \sum_{i=1}^{n} p_i z_i;\ u = \sum_{i=1}^{n} p_i u_i;\ t = \sum_{i=1}^{n} p_i t_i \quad (1)$$

where p_i is the mass fraction of each constituent present in the mixture and x_i, y_i, z, u_i, t_i represent respectively the number of carbon (C), hydrogen (H), oxygen (O), nitrogen (N) and sulphur (S) atoms of each constituent i in the mixture.

The molar mass is defined as the average value of the molar masses of its constituents.

$$M_{C_xH_yO_zN_uS_t} = \sum_{i=1}^{n} p_i M_i \quad (2)$$

M_i is the molecular weight (in g/mol) of molecule i in the fuel.

The following atomic molar masses in g/mol are considered:

$$M_C = 12;\ M_H = 1;\ M_O = 16;\ M_N = 14;\ M_S = 32$$

Method of determining the density of fuels. The density of each fuel at temperature T is determined experimentally by the densimeter method. This method is based on the standardised method NF T60 101 (Guillermic, 1980). As HFO has variable characteristics and given its very high share in consumption (86.5%), three (03) samples of different origins (Cotonou, Lomé and Abidjan) were submitted to the analysis. For DDO, only one sample from Cotonou was submitted for analysis (Fig. 2).

The equipment used consisted of: a 500 mL capacity graduated cylinder, a densimeter graduated from 0.950 to 1 for HFO and 0.800 to 0.900 for DDO, a HANNA Checktemp digital thermometer and a Potence.

The following equation is used to reduce the density at the measurement temperature T to the reference temperature 15 °C (Guillermic, 1980).

$$d_4^{15} = d_4^{T} + A\,(T - 15) \quad (3)$$

where: d_4^T is the density at temperature T, d_4^{15} is the density at 15 °C and A is the density correction factor.

To choose the coefficient A, we use the table given by Guillermic (1980).

The density at 15 °C of the HFO will be the average of the densities at 15 °C of the three samples.

Method for determining the lower heating value of fuels. The lower calorific value of fuels is estimated using the following equation (Sawerysyn, 1993).

$$LCV = HCV - w\,L_v \quad (4)$$

LCV: lower calorific value (in kJ/kg).
HCV: higher calorific value of the fuel (in kJ/kg).
w: Total mass of water in kg released by the combustion of 1 kg of raw fuel.
L_v: the latent heat of vaporisation of water (in kJ/kg).

Fig. 2 Photograph of HFO and DDO fuel

The HCV or gross heat of combustion of fuels consumed by thermal power plants is estimated using Eq. (5) as it gives a fairly high accuracy with a deviation of generally less than 1% (Speight, 2006).

$$HCV = 12{,}400 - 2{,}100\, d^2 \tag{5}$$

where: HCV is in cal/g and d is the specific gravity 60/60 °F or density at 15.6 °C.

The following data are considered: 1 kcal = 4.185 kJ; L_v = 2.51 MJ/kg at 0 °C.

In this study, the density at 15 °C of the fuels is used to estimate the HCV.

$$w = \frac{\%\,E}{100} + \frac{9\,\%\,H}{100} \tag{6}$$

% E is the overall moisture content of the raw fuel and % H is the hydrogen content of the fuel.

Determination of CO_2 emission factors for fuels. The CO_2 energy EF of fuel f for combustion is calculated using the formula used by Ramphull and Surroop (2017)

$$EF_{CO_2,f} = \frac{\%\,C_f \times FO \times 3{,}667}{LCV_f} \tag{7}$$

where: $EF_{CO_2,f}$ is the CO_2 energy emission factor of fuel f for combustion (in $kgCO_2/TJ$ or $kgCO_2/kW$ h etc.); $\%C_f$ is the carbon content (in kg carbon/kg fuel) and 3.667 is the ratio of the molar mass of CO_2 to that of carbon; FO is the oxidation factor of carbon; LCV_f is the net calorific value of fuel f (TJ/kg or kJ/kW h etc.) and $EF_{m,f}$ is the CO_2 mass emission factor of fuel f (in $kgCO_2$/kg fuel).

Equivalence: 1 kW h = 3600 kJ.

The CO_2 mass EF of each fuel is determined according to the following formula (ADEME, 2016):

$$EF_{m,f} = EF_{CO_2,f} \times LCV_f = \%\,C_f \times 3{,}667 \tag{8}$$

where: $EF_{m,f}$ is the CO_2 mass emission factor of fuel f (in $kgCO_2$/kg fuel f)

Methodology for quantifying carbon dioxide emissions. The following assumptions are applied: (i) the FO of carbon is assumed to be 100%; (ii) the results of the HFO and DDO analyses are valid for all fuels consumed by SONABEL's different thermal power plants.

The CO_2 emissions from combustion are evaluated using the following equation (CITEPA, 2020).

$$E_{CO_2} = \sum_{f=1}^{n} m_f \times LCV_f \times EF_{CO_2,f} \tag{9}$$

where: E_{CO_2} are the total CO_2 emissions from fuel combustion (in kg);
m_f is the quantity of fuel f (in kg).

Determination of the CO_2 emission factor of thermal power generation. The CO_2 EF of thermal power generation is assessed by dividing the amount of CO_2 emitted in 2018 by the thermal power generation of the same period.

$$EF_{th} = \frac{E_{CO_2}}{E_{é\,1,th}} \tag{10}$$

where EF_{th} is the CO_2 emission factor of thermal power generation in $kgCO_2/kW$ h and $E_{é\,1,th}$ is the power generation of thermal plants.

Determination of the CO_2 emission factor of the electricity generation mix. The CO_2 EF of SONABEL's electricity generation mix (taking into account all generation sources, i.e. thermal and renewable) is calculated by averaging the EFs of the different electricity generation sources. Hydroelectricity and solar photovoltaic do not emit CO_2 during the electricity generation phase; it comes that:

$$EF_{el} = EF_{th} \cdot \chi_{th} \tag{11}$$

where: EF_{el} is the CO_2 emission factor of the electricity generation mix (in $kgCO_2/kW$ h) in 2018; EF_{th} is the CO_2 emission factor of thermal generation (in $kgCO_2/kW$ h) and χ_{th} represent the share of thermal generation in the generation mix in 2018.

4 Results and Discussion

4.1 Constituents of HFO and DDO

Figures 3 and 4 show the constituents of HFO and DDO respectively.

The HFO consumed by power stations contains mainly cycloalkanes (or naphthenes with 89%) and also contains small amounts of organic nitrogen compounds (pyridine and pyrrole), sulphur contained in methylthiophene, and does not contain oxygen. The indeterminates in HFO are very high and this can be explained by several factors including the source of the fuel, as there are several varieties of heavy fuel oil. DDO contains 83.5% non-cyclic alkanes or paraffins and 15% cycloalkanes or naphthenes. It is also noticeable that S, N and O are not present in DDO. The indeterminates in DDO are less important compared to those in HFO. Previous studies have shown that petroleum hydrocarbons contain mainly paraffins, naphthenes, olefins and aromatic hydrocarbons (Gruyer et al., 2015). Guibet (1997) and Laxalde (2012) have suggested that crude oil and refinery products contain other elements such as sulphur, nitrogen, oxygen and some metals including nickel and vanadium in low or trace concentrations, especially in the heavier fractions.

4.2 Estimated Physicochemical Properties of the Fuels

Table 1 provides a summary of the results of the analyses of the fuels studied in terms of their physicochemical characteristics. The measured temperature of DDO is 29 °C and the average temperature of HFO is 28.7 °C.

Elemental mass composition. The estimated C, S and N content for HFO and the C content of DDO are in agreement with the elemental mass composition provided by Speight (2006) (C (83 to 87%); H (10 to 14%); N (0.1 to 2%), S (0.05 to 6%)) and also falls within the limits given by Guillermic (1980) (C (84 to 87%); H (10.5 to 14%); S (0.2 to 4%)). Only hydrogen is an exception for both fuels with a mass fraction above 14. Indeed, in the literature, the mass fraction of hydrogen in petroleum fuels is between 10 and 14% (Guillermic, 1980; Speight, 2006). This observed difference can be explained for two reasons: firstly, the indeterminate part is probably due to the origin of the oil and the refining conditions and secondly, the method of characterisation of the fuels.

The results of the work of Kornelius et al. (2022) on the analysis of liquid and gaseous fuels used in South Africa gave a C content of 85.93 ± 1.58% for HFO; and

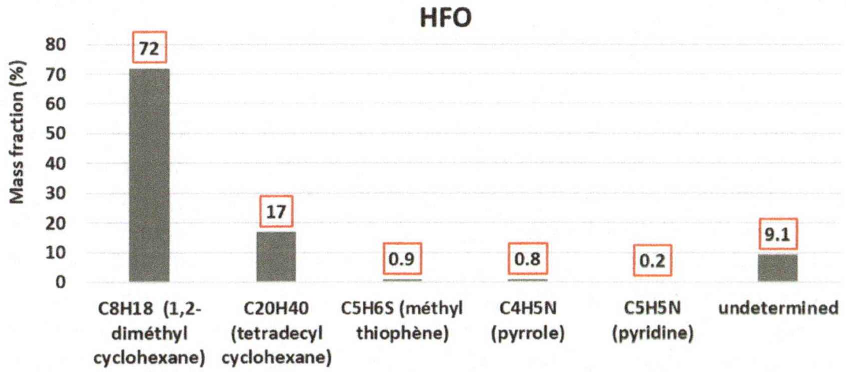

Fig. 3 Molecular composition of HFO

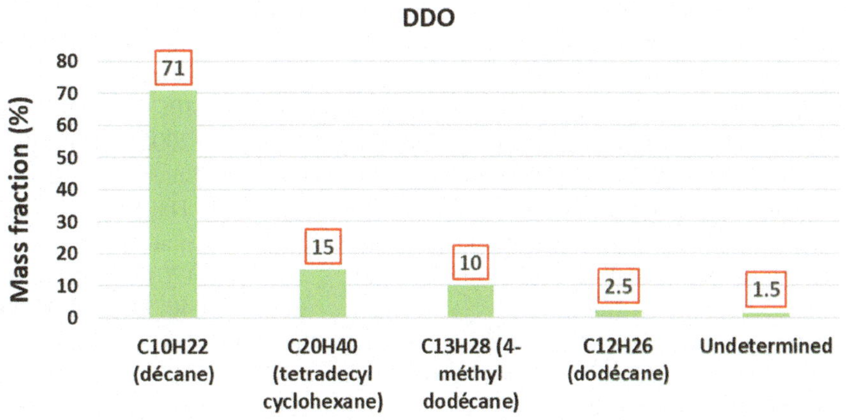

Fig. 4 Molecular composition of DDO

Table 1 Results of analyses of HFO and DDO delivered to Burkina Faso

Fuel characteristics	Symbol/notation	Units	HFO	DDO
Chemical formula	–	–	$C_{9.247}H_{19.864}N_{0.01}S_{0.009}$	$C_{11.7}H_{25.07}$
Carbon	C	%	84.54	84.85
Hydrogen	H	%	15.13	15.15
Sulphur	S	%	0.22	–
Nitrogen	N	%	0.11	–
Indeterminate	–	%	9.1	1.5
Density measured at the temperature T	d_4^T	–	0.956	0.827
Estimated density at 15 °C	d_4^{15}	–	0.964	0.837
Water content	E	%	0.05	ND
Estimated lower calorific value at 15 °C	LCV	kJ/kg	40,307.76	42,314.7
		kW h/kg	11.20	11.75

for diesel, it is 88.31±0.79% in the summer period and 87.00±0.38% in the winter period. There is a difference of less than 2% between the C content value estimated in this study for Burkinabe HFO and South African HFO. The C content of DDO consumed in Burkina Faso is 3.92% and 2.5% lower than that of South African diesel (in summer and winter respectively).

Density. According to national oil and gas company of Burkina Faso (named SONABHY), the density of fuel oil at 15.6 °C is in the range 0.920 – 0.990 kg/dm^3 according to the ASTM or AFNOR method (SONABHY, 2021). The density evaluated for the different HFOs is in the range given by SONABHY. According to Benin Control (2019), the density at 15 °C of heating oil (FOD or DDO) is between 0.820 and 0.930 kg/dm^3. For fuel oil 180 cSt, its density at 15 °C is between 0.920 and 0.995 kg/dm^3 and for fuel oil 380 cSt, its density at 15 °C is lower than 0.995 kg/dm^3 according to the AFNOR or ASTM method. The density assessed for HFO and DDO in this study is in accordance with SONABHY and Benin specifications.

The estimated result for the density of HFO is consistent with data from Wauquier (1994) and GIEC (2006a). Indeed, the density of heavy fuel oils is higher than 0.920 kg/L at 15 °C (Wauquier, 1994). For the IPCC, the density of heavy residual fuels, including those obtained by blending, always exceeds 0.90 (GIEC, 2006a).

Kornelius et al. (2022) in their work estimated a density of 0.994±0.12 kg/L for Heavy Fuel Oil (HFO) and 0.826±0.002 kg/L and 0.825±0.002 kg/L for diesel used in South Africa in summer and winter respectively. The density of DDO is 1.3% higher than that of the diesel evaluated by Kornelius et al. (2022).

Water content. The water content evaluated for the HFO consumed by SONABEL's thermal power plants is 0.05% and therefore it contains a very low water content. For DDO, we considered its water content to be zero because, on the one hand, it is a cleaner fuel than HFO and even closer to diesel, and on the other hand, the visual aspect shows that it does not have traces of water.

According to SONABHY (2021), the water content of fuel oil is a maximum of 0.50% by volume. According to Benin Control (2019), the technical specification for the water content of heating oil (FOD or DDO) is 200 mg/kg (i.e. 0.0002%) maximum. For 380 cSt fuel oil, the water content by distillation is 1.0% by mass maximum.

Lower heating value of fuels. The total mass of water released by the combustion of 1 kg of HFO is 1.3622 kg, and for 1 kg of DDO it is 1.3635 kg.

The LCV of HFO is slightly lower than the IPCC default for residual fuel oil, which is 40,400 kJ/kg (GIEC, 2006a). As shown in Table 1, the LCV of HFO is within the 95% confidence interval of the IPCC default for residual fuel oils (39.8 and 41.7 TJ/Gg). The HCV and LCV of fuel oil according to SONABHY's technical specifications are respectively 43,000 kJ/kg minimum and 40,000 kJ/kg minimum (SONABHY, 2021). Ramphull and Surroop (2017) estimated an LCV value of 40.6 MJ/kg for HFO 380 cSt and 41.1 MJ/kg for HFO 180 cSt. The difference between the LCV of 380 cSt fuel oil and that obtained in our study is less than 1%. For HFO 180 cSt, the difference is less than 2%. The LCV of DDO is slightly lower than the IPCC default LCV of 43,000 kJ/kg for diesel/diesel (GIEC, 2006a). Thus, the DDO LCV is within the 95% confidence interval of the default LCV for diesel (41.4–43.3 MJ/kg). The LCV of DDO (42.31 MJ/kg) is very close to that estimated for diesel (42.3 MJ/kg) by Ramphull and Surroop (2017). The LCV values for HFO and DDO are between 40,000 and 45,000 kJ/kg, a range given for a large number of hydrocarbons and petroleum fractions (Wauquier, 1994). According to CITEPA (2022), the PCI used in France is 40 MJ/kg and 42.6 MJ/kg for heavy commercial

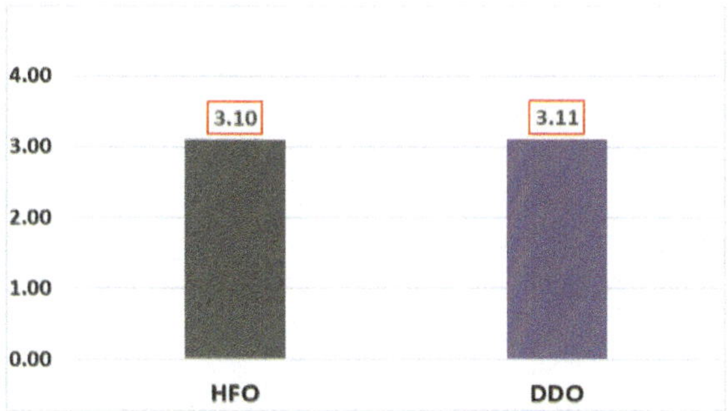

Fig. 5 CO_2 mass emission factors for HFO and DDO in kg CO_2/ kg fuel

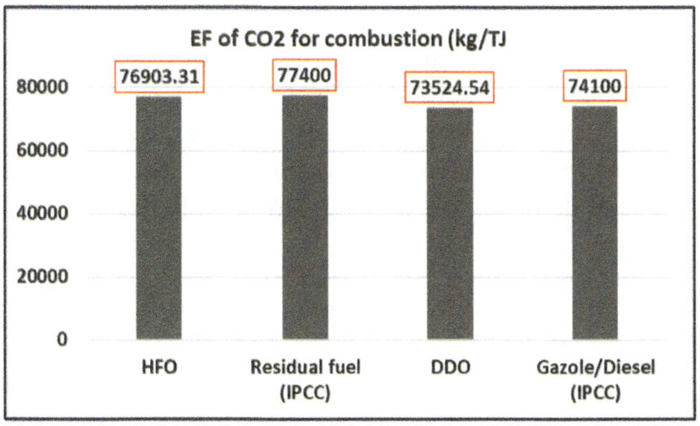

Fig. 6 CO_2 emission factors for HFO and DDO combustion and IPCC default emission factors for residual fuel oils and gas oil/diesel in kg/TJ

fuel oil and domestic fuel oil/gasoil respectively. The LCV evaluated for DDO is respectively close to that evaluated for diesel by Kornelius et al. Indeed, the latter estimated an LCV of 43.05 ± 0.07 MJ/kg in the summer period and 42.99 ± 0.033 MJ/kg in the winter period for diesel used in South Africa.

CO_2 mass and energy emission factors for fuel combustion. Figures 5 show the CO_2 mass emission factors for HFO and DDO respectively.

The CO_2 mass EFs assessed for the two fuels are respectively lower than the CITEPA (2022) data for commercial heavy fuel oil (3.12 kg CO_2/kg fuel) and diesel/heating oil (3.17 kg CO_2/kg fuel); but overall, there are small differences. Kornelius et al. estimated a volume EF of 3.124 kg/L for HFO used in South Africa. Converting this volume EF using the density of the HFO they estimated (0.994 kg/L) gives a mass EF of 3.14 kg CO_2/kg HFO. We find that the CO_2 mass EF of the HFO consumed by SONABEL's thermal power plants is 1.27% lower than that of the HFO used in South Africa.

Figure 6 shows the CO_2 EFs assessed for HFO and DDO combustion and the IPCC default EFs for residual fuel oils and gas oil/diesel on a net calorific basis (GIEC, 2006b).

The CO_2 energy EF for HFO combustion and that for DDO evaluated in this study are slightly lower than the IPCC default EFs for residual fuel oils and gas oil/diesel respectively. The same discrepancy is observed between DDO and gas oil/diesel. In addition, the CO_2 EF of HFO and DDO are found to be within the 95% confidence interval provided by the IPCC for residual fuel oils (75,500–78,800 kg/TJ) and for gas oil/diesel (72,600–74,800 kg/TJ) respectively. For the IPCC, when a specific national EF falls within the 95% confidence interval, this EF can be considered consistent with the default value (GIEC, 2006b). Similar studies have shown that the EF for HFO 180 cSt; HFO 380 cSt and HFO 324 cSt are 74,344 kg/TJ, 78,028 kg/TJ and 75,333 kg/TJ respectively (Ramphull & Surroop, 2017). The EF assessed for HFO is 3.33%; 1.44% and 2.04% lower than that of HFO 180 cSt, HFO 380 cSt and HFO 324 cSt assessed by Ramphull and Surroop (2017) respectively.

The analysis of the results obtained for the CO_2 energy EFs shows that HFO combustion emits more CO_2 than DDO combustion at equal energy production. According to Soleille (2004), the combustion of heavy fuel oil emits more CO_2, at equal energy production, than that of natural gas or lighter petroleum products (gasoline, diesel, domestic fuel oil) but less than that of coal or petroleum coke.

4.3 Contribution of the Burkinabe Electricity Generation Sector to CO_2 Emissions

In 2018, total CO_2 emissions attributable to fuel combustion amounted to 579.83 Gg, i.e. 501.31 Gg due to the combustion of HFO and 78.52 Gg due to DDO. This represents 86.5% from HFO and 13.5% from DDO. The Burkinabe electricity production sector is a sector that contributes significantly to CO_2 emissions through the use of fossil fuels, in particular petroleum products.

The environmental management of functional units and environmental assessments related to SONABEL's equipment projects have estimated in terms of pollution statistics 583.94 Gg of CO_2 for the period 2018 (SONABEL, 2018). The difference is about 0.7% compared to the one estimated in this study and this difference is due to the fact that the CO_2 emissions estimated in this work use country-specific EFs and not default factors.

According to the final results of the fifth General Census of Population and Housing (RGPH) of Burkina Faso, its population is estimated to be 20,505,155 in 2019 (INSD,

2022). Thus, the CO_2 emissions caused by SONABEL's electricity generation amounted to 28.28 kg per capita in 2019.

4.4 CO_2 Emission Factor of Thermal Power Generation

The CO_2 EF of SONABEL's thermal power generation is estimated at 0.663 kg/kW h LCV for the year 2018 in this study. This value is quite high and it clearly shows that fossil fuels are the main emitters of CO_2.

According to ADEME (2016), the EF of electricity production for the steam-oil plant is 704 geqCO_2/kW h. Fossil fuel technologies have higher emission factors (that of natural gas is almost half that of coal, and about two thirds of the emission factor of fuel oil) (Spadaro et al., 2000). According to Afilal et al. (2019), electricity generated from fuel oil has an EF of 730 gCO_2 eq/kW h. The discrepancies found are due to the fact that the value assessed in this work does not include other GHGs but CO_2 only.

4.5 CO_2 Emission Factor of the Electricity Generation Mix

The CO_2 EF of the SONABEL power generation mix is estimated at 0.569 kg/kW h LCV for 2018. This value is quite high because of the mode of electricity production which is mostly thermal. It highlights the largely carbonated nature of the electricity produced in Burkina Faso.

Sanogo et al. (2022) assessed the EF of Burkina Faso's electricity network using SONABEL's production data according to the UNFCCC methodology. The results show that the EF of the simple operating margin of the Burkina Faso power system varies from 0.552 to 0.624 tCO_2/MW h between 2015 and 2019, and is 0.584 tCO_2/MW h for the year 2019. It can be seen that the EF estimated in this study for the year 2018 is within the range given by Sanogo et al. (2022) and is lower than 0.584 tCO_2/MW h with a gap of 2.36%. This gap is due to the fact that our study is limited to the generation phase and not to the whole electricity network which takes into account emissions due to electricity exchanges at interconnections (electricity imports).

According to DEFRA/DECC (2014), the international emissions conversion EF for electricity used on the Creative IG Tools for the purpose of calculating an organisation's carbon footprint is 0.596 kgCO_2/kW h for regions including Africa and for countries that do not have their own emissions conversion factor. We note that the CO_2 EF value of the power generation mix assessed in this study is about 4.5% lower than the 0.596 kgCO_2/kW h value.

In 2018, the global average carbon intensity of electricity produced was 0.475 kgCO_2/kW h (CITEPA, 2019). It can therefore be seen that the CO_2 EF of the Burkinabe electricity generation mix (SONABEL) is higher than the global average. In the West African sub-region, the carbon intensity of electricity (in teqCO_2/MW h) in Côte d'Ivoire, Niger and Benin are respectively 0.385, 0.630 and 0.684 (BP, 2022).

4.6 Assessment of the Contribution of Renewables to the Reduction of CO_2 Emissions from the Electricity Generation Sector

In 2018, the share of renewable energy (RE) in SONABEL's electricity production is 145.5 GW h (i.e. 14.25% of total production). The CO_2 emissions avoided in 2018 are evaluated by considering that the electricity production from RE sources is generated by thermal power plants with an EF_{th} of 0.663 kg/kW h (value calculated in this study).

$$E_{CO_{2Ev}} = E_{ER} \times EF_{th} \tag{12}$$

$E_{CO_{2Ev}}$: CO_2 emissions avoided (in Gg);

E_{ER}: Electricity production from RE sources (in GW h).

We evaluate the avoided emissions at nearly 96.5 Gg of CO_2 or 16.7% of the total CO_2 emissions attributable to SONABEL's electricity generation in 2018.

According to CITEPA (2019), an increased use of RE in 2018 avoided 215 Mt of CO_2 emissions, mainly in the power generation sector in Europe and China. The analysis of the results shows that the reduction of anthropogenic CO_2 emissions in the power generation sector can be achieved by increasing the share of RE in the electricity mix. Thus, electricity production from RE sources is a means of climate change mitigation. According to an IEA report, RE will transform the global electricity mix by 2027, overtaking coal as the leading source of electricity generation (IEA, 2022).

5 Conclusion

The results obtained in the present study show that the CO_2 EFs for HFO and DDO combustion are slightly lower than those provided by the IPCC for residual fuel oils and gas oil/diesel respectively. However, the CO_2 EF for HFO and DDO are respectively within the 95% confidence interval provided by the IPCC for residual fuel oils and for gas oil/

diesel. The EFs evaluated in this study take into account the physicochemical characteristics of the fuels consumed by thermal power plants. These EFs are country-specific and therefore better adapted to the national context for the quantification of CO_2 emissions from electricity generation, particularly in the context of national GHG emission inventories under the UNFCCC. The CO_2 EF assessed for thermal power generation is quite high and clearly indicates that fossil fuels, in particular oil products, are the main CO_2 emitters. The CO_2 EF of SONABEL's electricity production mix assessed for 2018 is relatively high and highlights the largely carbon-based nature of electricity produced in Burkina Faso (mainly thermal electricity production). Burkina Faso has enormous renewable energy potential, particularly solar energy. Increasing the share of RE in the production mix will, on the one hand, greatly reduce emissions from the sector and, on the other hand, solve the electricity deficit.

Acknowledgements This work was conducted within the Laboratoire de Physique et de Chimie de l'Environnement (LPCE). The objectives were achieved thanks to the support and collaboration of several people. We reiterate our sincere thanks to the managers of SONABEL.

References

ADEME. (2016). Documentation des facteurs d'émissions de la Base Carbone® 11.5.0 (p. 287).

Afilal, M. E., Abbadin, Laiche, H., & Elfarh, A. (2019). Le mix énergétique et les GES au Maroc : Évaluation des Scenarios et mesures prévues," Revue Espace Géographique et Société Marocaine (Vol. no. 26, pp. 273–290).

Benin Control. (2019). Receveur des douanes. 08 Août 2019. Retrieved Février 25, from https://benincontrol.com.

BP Statistical Review of World Energy. (2022). BP Statistical Review of World Energy. Retrieved Février 23, 2020, from https://www.bp.com/content/dam/bp/business-sites/en/global/corporate/pdfs/energy-economics/statistical-review/bp-stats-review-2022-full-report.pdf.

CDKN. (2022). (Alliance pour le Climat et le Développement) et ACDI (l'Initiative africaine pour le climat et le développement). LE SIXIÈME RAPPORT D'ÉVALUATION DU GIEC, Impacts, options d'adaptation et domaines d'investissement pour une Afrique de l'Ouest résiliente au changement climatique.

CITEPA. (2019). Emissions mondiales de CO_2: 2e année de hausse après trois ans de stabilisation. https://www.citepa.org/fr/2019_04_a1/.

CITEPA. (2020). Organisation et méthodes des inventaires nationaux des émissions atmosphériques en France, OMINEA—17ème édition. https://www.citepa.org/wp-content/uploads/publications/ominea/Citepa_Ominea_ed2020.pdf.

CITEPA. (2022). Facteurs d'émission CO_2 et pouvoirs calorifiques inférieurs (PCI) nationaux (valeurs par défaut) par type de combustible, France.

DEFRA/DECC (The Department for Environment, Food and Rural Affairs and Department of Energy and Climate Change). (2014). List of the international emission conversion factors for electricity, Guidelines for DEFRA/DECC's GHG.

GIEC. (2006a). Lignes directrices 2006 du GIEC pour les inventaires nationaux de gaz à effet de serre. Volume 2 : Énergie, Chapitre 1 : Introduction.

GIEC. (2006b). Lignes directrices 2006 du GIEC pour les inventaires nationaux des gaz à effet de serre. Volume 2 Energie, Chapitre 2 : Combustion stationnaire.

GIEC. (2022). Synthèse vulgarisée du résumé aux décideurs du groupe de travail III de l'AR6 – The Shifters.

GIEC. (2021). Changement climatique 2021—Les bases scientifiques physiques—Résumé à l'intention des décideurs. Contribution du Groupe de travail I au sixième Rapport d'évaluation du Groupe d'experts intergouvernemental sur l'évolution du climat.

Gruyer, N., Groleau, P.-É., Triffault-Bouchet, G., Ouellet, A. & Dupont, F. (2015). Hydrocarbures pétroliers : caractéristiques, devenir et criminalistique environnementale –Études GENV222 et GENV23 (p. 41 p. et annexes). Centre d'Expertise en Analyse Environnementale du Québec.

Guibet, J.-C. (1997). *Carburants et moteurs-Technologies-Energie-Environnement, Tome 1*. Publication de l'Institut Français du Pétrole.

Guillermic, A. (1980). Le chauffage par les combustibles liquides. In TECHNIP (Ed.), *Caractéristiques des combustibles, Tome 1* (pp. 17–243). Institut Français du pétrole.

IEA (International Energy Agency). (2019a). Global energy and CO_2 status report 2018: The latest trends in energy and emissions in 2018.

IEA (International Energy Agency). (2019b). Africa Energy Outlook 2019 | Rapport Spécial.

IEA (International Energy Agency). (2022). *Renewables 2022: Analysis and forecast to 2027*.

INSD (Institut National de la statistique et de la démographie). (2022). Cinquième Recensement Général de la Population et de l'Habitation du Burkina Faso-Synthèse des résultats définitifs. Burkina Faso.

Kornelius, G., Forbes, P., Fischer, T., & Govender, M. (2022). Tier 2 greenhouse gas emission factors for South African liquid and gaseous fuels. *Clean Air Journal, 32*(2), 2.

Laxalde, J. (2012). Analyse des produits lourds du pétrole par spectroscopie vibrationnelle. Chimie théorique et/ou physique (pp. 24, 25 et 26. Thèse de Doctorat, Université des Sciences et Technologie de Lille.

ME (Ministère de l'Energie du Burkina Faso). (2018a). Annuaire Statistique. Burkina Faso.

ME (Ministère de l'Energie du Burkina Faso). (2018b). Tableau de bord, Burkina Faso.

Nations Unies. (2015). Accord De Paris.

PRBA. (2021). Premier rapport biennal actualisé du Burkina Faso sous la CCNUCC.

Ramphull, M., & Surroop, D. (2017). Greenhouse gas emission factor for the energy sector in Mauritius. *Journal of Environmental Chemical Engineering*, 5994–6000.

Sanogo, O., Haro, K., Compaoré, R. C. M., Kouraogo, R. P. & Palm, K. (2022). Evolution du Facteur d'émission électrique des réseaux électriques des pays de l'Afrique Subsaharienne : cas du Burkina Faso (pp. 232–243). Revue science et technique.

Sawerysyn, J.-P. (1993). Les pouvoirs calorifiques. In Bulletin de l'Union des physiciens (Vol. 87, pp. 401–411).

Soleille, S. (2004). L'industrie du raffinage et le devenir des fiouls lourds, Rapport final.

SONABEL (Société Nationale Burkinabé d'Electricité). (2018). Rapport d'activité.

SONABHY (Société Nationale Burkinabè d'Hydrocarbure). (2021). Télécopie N° 21 / 00020 MCIA/SONABHY/DG/DAS, Fourniture de produits pétroliers, 22 Janvier 2021. [Accessed 27 Février 2023].

Spadaro, J. V., Langlois, L., & Hamilton, B. (2000). Emissions de gaz à effet de serre provenant des chaînes de production d'électricité, Evaluer la différence (pp. 19–24). AIEA BULLETIN 42/2/2000.

Speight, J. G. (2006). *The chemistry and technology of petroleum* (4th ed., p. 955).

Wauquier, J.-P. (1994). Le raffinage du pétrole. Pétrole brut, Produits pétroliers, Schémas de fabrication, Caractéristiques des produits pétroliers utilisés à des fins énergétiques (carburants - combustibles) - Tome 1, Publication de l'Institut Français du Pétrole, TECHNIP.

Organic and Elemental Carbon in $PM_{2.5}$ from an Urban Residential Area of Lagos, Nigeria

Oluwabamise Lekan Faboya, Khanneh Wadinga Fomba and Hartmut Herrmann

Abstract

Exposure to ambient organic matter can influence human health and is a source of poor quality contributing to the formation of photochemical smog. In this study, organic carbon (OC) and elemental carbon (EC) concentrations were determined in $PM_{2.5}$ particle fraction during the height of the rainy season in Lagos, Nigeria (June–July 2021). The daily average $PM_{2.5}$ mass concentration (27 µg m^3) exceeded the World Health Organization (WHO) limit of 15 µg m^{-3}. The 24-h average EC and OC concentrations values were 0.8 and 2.2 µg m^{-3}, respectively, with both OC and EC contributing 11% of the total $PM_{2.5}$ mass. A good correlation existed between OC and EC, suggesting the same origin. Considering the OC/EC ratio in the range of 2 to 5.1 µg m^{-3}, the most likely common sources of carbonaceous aerosols in the investigated residential area are vehicular emissions, motorbikes, and generator fumes. The observed relatively high $PM_{2.5}$ mass and the proportion of carbonaceous components in the total $PM_{2.5}$ mass in the residential area indicate that the air quality in Lagos is already at risk and needs immediate monitoring.

Keywords

Air Quality · Urban Aerosol · Carbonaceous Species · OC/EC

1 Introduction

Elemental carbon (EC) and organic carbon (OC) are carbon-constituent species that are components of inhalable particulate matter with a diameter of 2.5 µm or smaller ($PM_{2.5}$) (Winquist et al., 2015). Increased rates of cardiovascular morbidity and death have been tied to exposure to elevated concentrations of these carbon-containing species (Samara et al., 2014). The OC includes several classes of pollutants originating from both primary and secondary sources. EC is emitted through anthropogenic combustion activities. The conversion processes of volatile precursors from gases into particles in the atmosphere result in the production of secondary organic carbon (SOC), while the various combustion processes often lead to the direct emission of primary organic carbon (POC) into the atmosphere.

Lagos, the former capital of Nigeria, is the largest city in West Africa with an estimated 21 million inhabitants (Alani et al., 2019). The city is prone to different sources of aerosol contamination. It has continued to be the business/industrial center of Nigeria, where a wide range of events take place that may negatively impact the air quality of the city. Other factors that may contribute to the air pollution level in Lagos may include emissions from woodstoves, waste burning, soil dust, traffic pollution from extremely old and rickety cars and motorbikes, and toxic fumes emanating from generators due to unstable electricity supply. The present study, therefore, investigated $PM_{2.5}$ mass concentration, organic carbon, and elemental carbon in $PM_{2.5}$ samples collected from a residential area in Lagos, Nigeria. The study is expected to improve on the dearth of information on the carbonaceous aerosol from the city.

O. L. Faboya · K. W. Fomba · H. Herrmann (✉)
Department of Chemistry, Federal University Oye-Ekiti, Oye-Ekiti, Ekiti State, Nigeria
e-mail: herrmann@tropos.de

O. L. Faboya
Atmospheric Chemistry Department (ACD), Leibniz Institute for Tropospheric Research (TROPOS), Permoserstraße 15, 04318 Leipzig, Germany

K. W. Fomba et al. (eds.), *Advances in Air Quality Research in Africa*, Advances in Science, Technology & Innovation, https://doi.org/10.1007/978-3-031-53525-3_2

2 Materials and Methods

2.1 Site Description and Sample Collection

A total of 36 $PM_{2.5}$ samples (24-h samples) and blanks (4) were collected from a residential area within the premises of 3/5 Morenike Carena Close, Orile Agege, Lagos (6.635008N and 3.302543E) on high purity quartz-fiber filters ($\Phi = 47$ mm) using a Low-Vol "Gent" stacked $PM_{2.5}$ air sampler (Maenhaut et al., 1994). The equipment was run at a steady flow rate of 18 L/min and installed at 4.65 m above the ground level to prevent obstruction. Sampling was carried out between June and July 2021, at the height of the rainy season in Nigeria.

2.2 Gravimetric and OC/EC Analysis

The $PM_{2.5}$ mass concentration was measured using a microbalance (Mod. AT261 Delta Range, Mettler-Toledo). The filters were conditioned for 72 h before and after the sampling at a steady temperature (20 ± 1 °C) and relative humidity (50 ± 5%). To determine the $PM_{2.5}$ mass concentration, the weight differential of the filters (weight after sampling minus weight before sampling) was divided by the overall sampling volume.

OC and EC contents were measured using a thermal-optical method reported by Van Pinxteren et al. (2015), following the established EUSAAR2 protocol described by Cavalli et al. (2010). The OM, POC, and SOC values were derived according to the following formulas (Huang et al., 2012; Turpin & Lim, 2001):

$$OM = 2.1 \times OC \quad (1)$$

$$POC = (OC/EC)_{min} \times EC \quad (2)$$

$$SOC = OC - POC \quad (3)$$

where the lowest OC/EC ratio found during the campaign period is denoted by $(OC/EC)_{min}$.

3 Results and Discussion

3.1 $PM_{2.5}$ Mass

The $PM_{2.5}$ mass concentration and the carbonaceous components in the $PM_{2.5}$ are illustrated in Fig. 1. The daily variation in $PM_{2.5}$ mass, OC, and EC are also shown in Fig. 2. The 24-h $PM_{2.5}$ mass concentrations were in the range of 8.7 and 56.5 µg m^{-3} with an average value of 27 µg m^{-3}, exceeding the daily 15 µg m^{-3} threshold limit recommended by World Health Organization (WHO, 2021; Turpin & Lim, 2001). The average mass concentration was higher than the previous values reported by authors from Lagos during a similar sampling period (rainy season) (Alani et al., 2019 and ref. therein). However, the value was comparable to the value reported by Owoade et al. (2013) from a residential area in Lagos (28 µg m^{-3}). The daily fluctuations in $PM_{2.5}$ mass concentrations may be attributed to anthropogenic activities and/or the prevailing

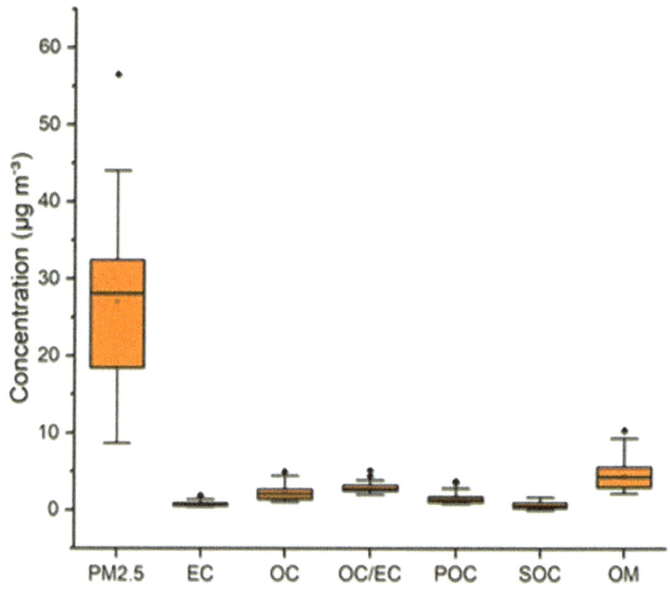

Fig. 1 $PM_{2.5}$ mass concentration and other carbonaceous constituents (µg/m³) in $PM_{2.5}$

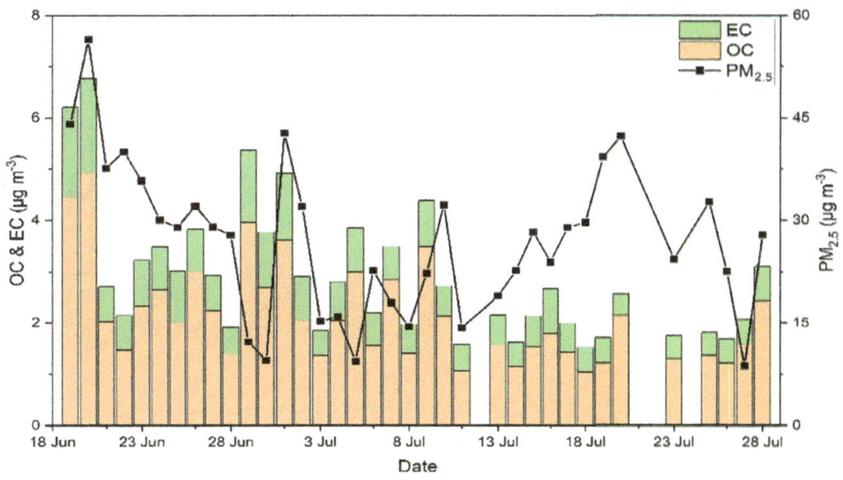

Fig. 2 Daily distribution of $PM_{2.5}$ mass, OC, and EC levels during the campaign period

meteorological conditions during sampling period. The observed average $PM_{2.5}$ mass concentration indicated that airborne PM pollution is already being witnessed in Lagos.

3.2 Carbonaceous Species

As shown in Fig. 1, the OM contents were in the range of 2.2 and 10.3 µg m^{-3} (av. 4.5 µg m^{-3}). Organic matter constituted 17% of the entire $PM_{2.5}$ mass on average. The enhanced OM content in $PM_{2.5}$ at the study site could be related to local emissions from anthropogenic sources. As also presented in Fig. 1, the OC and EC concentrations were between 1.0 and 4.9 µg m^{-3} (av. 2.2 µg m^{-3}) and 0.4 and 1.8 µg m^{-3} (av. 0.8 µg m^{-3}), respectively. Both OC and EC accounted for 11% of the total $PM_{2.5}$ mass on average, indicating that they were significant contributors to $PM_{2.5}$ mass and their sources therefore need to be controlled. The daily fluctuations in the OC and EC levels may equally be ascribed to anthropogenic activities and/or the prevailing meteorological conditions during sampling period (Fig. 2). As shown in Fig. 3a, strong association existed between OC and EC. This showed that most of the OC was produced daily as primary aerosol together with EC. This implied a shared origin of EC and OC, including the precursors that led to the formation of SOC. There was a strong impact of the primary combustion sources on the production of SOC during the sampling, as indicated by the strong positive bimodal correlation ($r^2=0.76$ and 0.40) between SOC and EC (Fig. 3b). It also implied two distinct relationships between the two carbonaceous species, like sources or atmospheric processes.

The OC/EC ratios were between 2 to 5.1 µg m$-^{3}$ (av. 2.9 µg m^{-3}) (Fig. 1). Hildemann et al. (1991) and Cao et al. (2005) determined 2.2 µg m$-^{3}$ and 4.1 µg m^{-3} for light-duty gasoline vehicles and vehicle exhaust, respectively. Thus, the plausible sources of $PM_{2.5}$ emission at this site were most likely car exhausts, motorbikes, and generator exhausts. OC/EC value higher than 2 is indicative of the SOC formation in the atmosphere (Javed et al., 2019), and exposure to secondary organic aerosols can have a negative impact on human health.

4 Conclusion

In this work, the $PM_{2.5}$ mass concentrations and carbonaceous components were determined in $PM_{2.5}$ filter samples collected during the height of the rainy season in Lagos, Nigeria. The average $PM_{2.5}$ mass concentration was almost twice higher than the limit established by the WHO. Organic matter contributed 17% of the total $PM_{2.5}$ mass, which may be attributed to local anthropogenic emissions. The 24-h average OC and EC concentration values were 2.2 and 0.8 µg m^{-3}, respectively. A good correlation between OC and EC suggested a common source. OC and EC accounted for 11% of the total $PM_{2.5}$ mass, indicating that they contributed significantly to $PM_{2.5}$ mass and therefore their sources need to be controlled. Based on the OC/EC ratios, the most likely prevalent sources of carbonaceous species in the residential area were vehicular emissions, motorbikes, and generator fumes.

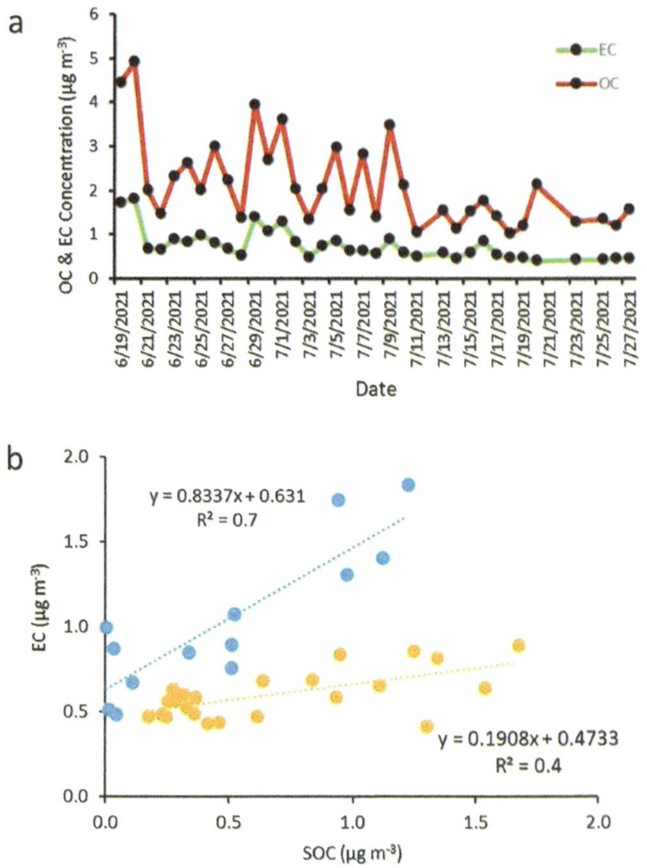

Fig. 3 **a** Temporal variation of OC and EC showing good correlation between the two parameters, and **b** EC versus SOC showing the relationship between the parameters

Acknowledgements The authors express their gratitude to the Alexander von Humboldt Foundation for making this research possible by awarding the first author (OLF) a postdoctoral research fellowship in Germany.

References

Alani, R. A., Ayejuyo, O. O., Akinrinade, O. E., Badmus, G. O., Festus, C. J., Ogunnaike, B. A., & Alo, B. I. (2019). The level $PM_{2.5}$ and the elemental compositions of some potential receptor locations in Lagos, Nigeria. *Air Quality, Atmosphere & Health, 12*, 1251–1258.

Cao, J. J., Wu, F., Chow, J. C., Lee, S. C., Li, Y., Chen, S. W., An, Z. S., Fung, K. K., Watson, J. G., Zhu, C. S., & Liu, S. X. (2005). Characterization and source apportionment of atmospheric organic and elemental carbon during fall and winter of 2003 in Xi'an China. *Atmospheric Chemistry and Physics, 5*, 3127–3137.

Cavalli, F., Viana, M., Yttri, K. E., Genberg, J., & Putaud, J. P. (2010). Toward a standardised thermal-optical protocol for measuring atmospheric organic and elemental carbon: The EUSAAR protocol. *Atmospheric Measurement Techniques, 3*(1), 79–89.

Hildemann, L. M., Markowski, G. R., & Cass, G. R. (1991). Chemical composition of emissions from urban sources of fine organic aerosol. *Environmental Science and Technology, 25*, 744–759.

Huang, H., Ho, K. F., Lee, S. C., Tsang, P. K., Ho, S. S. H., Zou, C. W., Zou, S. C., Cao, J. J., & Xu, H. M. (2012). Characteristics of carbonaceous aerosol in $PM_{2.5}$: Pearl Delta River region China. *Atmosheric Research, 104–105*, 227–236.

Javed, W., Iakovides, M., Euripides, G. S., Wolfson, J. M., Koutrakis, P., & Guo, B. (2019). Concentrations of aliphatic and polycyclic aromatic hydrocarbons in ambient $PM_{2.5}$ and PM_{10} particulates in Doha, Qatar. *Journal of the Air & Waste Management Association, 69*(2), 162–177.

Maenhaut, W., Francois F., & Cafmeyer, J. (1994). The 'Gent' stacked filter unit sampler for the collection of atmospheric aerosols in two size fractions: Description and Instructions for installation and use. In *Applied research on air pollution using nuclear-related technique* (Vol. 19, pp. 249–263). IAEA Report NAHRES. Vienna.

Owoade, O. K., Fawole, O. G., Olise, F. S., Ogundele, L. T., Olaniyi, H. B., Almeida, M. S., Ho, M.-D., & Hopke, P. K. (2013). Characterization and source identification of airborne particulate loadings at receptor siteclasses of Lagos Mega-City, Nigeria. *Journal of the Air and Waste Management Association, 63*(9), 1026–1035.

Samara, C., Voutsa, D., Kouras, A., Eleftheriadis, K., Maggos, T., Saraga, D., & Petrakaki, M. (2014). Organic and elemental carbon associated to PM_{10} and $PM_{2.5}$ at urban sites of northern Greece. *Environmental Science and Pollution Research, 21*, 1769–1785.

Turpin, B. J., & Lim, H. J. (2001). Species contributions to $PM_{2.5}$ mass concentrations: revisiting common assumptions for estimating organic mass'. *Aerosol Science and Technology, 35*, 602–610.

Van Pinxteren, M., Fiedler, B., van Pinxteren, D., Iinuma, Y., Körtzinger, A., & Herrmann, H. (2015). Chemical characterization of sub-micrometer aerosol particles in the tropical Atlantic Ocean: Marine and biomass burning influences. *Journal of Atmospheric Chemistry, 72*, 105–125.

Winquist, A., Schauer, J. J., Turner, J. R., Klein, M., & Sarnat, S. E. (2015). Impact of ambient fine particulate matter carbon measurement methods on observed associations with acute cardiorespiratory morbidity. *Journal of Exposure Science and Environmental Epidemiology, 25*, 215–221.

WHO (2021).*WHO global air quality guidelines. Particulate matter ($PM_{2.5}$ and PM_{10}), ozone, nitrogen dioxide, sulfur dioxide and carbon monoxide*. Geneva: World Health Organization; Licence: CC BY-NC-SA 3.0 IGO.

Occurrence of Volatile Halogenated Hydrocarbons in the Atmosphere Around e-Waste Sites in Lagos City, Nigeria

Emmanuel Gbenga Olumayede, Oluwabamise Lekan Faboya, Khanneh Wadinga Fomba and Chuckuembe C. Ojiodu

Abstract

Informal electrical and electronic waste (e-waste) recycling in Nigeria is threatening the health of the people, through the release of toxic chemicals (dioxin, flame retardants, and volatile halogenated hydrocarbons) into the air. In this work, air samples were collected from four locations in Lagos, Nigeria between March and July 2019 using an active sampling method with a low volume sampling pump (Acuro, Drager, Lubeck, Germany). The tube contains Chromosorb 106 as an adsorbent. The air samples were collected at the height of 1.5 m from the ground surface so as to reflect the human breathing zone. After sampling, the halogenated hydrocarbons were desorbed using carbon disulphide prior to gas chromatography-mass spectrometry (GC-MS) analyses. The result showed that 15 volatile halogenated hydrocarbons (VHHs) were detected in the atmosphere of Lagos City; including four (4) chlorofluorocarbon, two (2) hydrochlorofluorocarbons, and six (6) halogenated alkanes, 1 chloroalkene, and two (2) Chlorinated benzenes. The average concentration of the total VHHs ranged from 25.79 to 38.50 μg/m^3 with the lowest in the dry season month of March and the highest found in the wet season month of July. The occurrence of these chemicals in the atmosphere calls for monitoring due to their potentially damaging effects on human health.

Keywords

Electronic waste · Volatile halogenated hydrocarbons · Air quality · Lagos

1 Introduction

Lagos, Nigeria, is the hub of large-scale e-waste in the country. According to Lagos State Government data, the city features above 17.5 million inhabitants making it one of the biggest cities—and the fast-growing in the world (Nnorom & Osibanjo, 2008). In the last ten years, the importation of used electronics has assumed an alarming dimension in the country, especially considering the volume of un-reusable goods (wastes) destined for recycling and exported into the country in the name of reuse (Lin et al., 2021; Ogungbuyi et al., 2012). Unfortunately, approximately 30% of second-hand imports were estimated to be non-functioning (therefore need to be declared as e-waste): half of this amount is repaired locally and sold to consumers and the other half is unrepairable. Despite the identification of Lagos as a hotspot for the disposal of e-waste, there is a dearth of scientific studies on the health and safety impacts of refurbishing e-waste processing in Lagos. Furthermore, information on atmospheric levels of halogenated hydrocarbons and exposure of workers and nearby communities around the recycling sites are scanty. The main aim of this study was to assess the occurrence and concentrations of volatile halogenated hydrocarbons in ambient air around the e-waste processing sites in Lagos. The variation and characteristics of ambient volatile halogenated hydrocarbons (VHHs) in the city of Lagos were also examined.

E. G. Olumayede (✉) · O. L. Faboya
Department of Chemistry, Federal University Oye-Ekiti, Oye-Ekiti, Ekiti State, Nigeria
e-mail: emmanuel.olumayede@fuoye.edu.ng

O. L. Faboya · K. W. Fomba
Atmospheric Chemistry Department (ACD), Leibniz Institute for Tropospheric Research (TROPOS), Permoserstraße 15, 04318 Leipzig, Germany

C. C. Ojiodu
Department of Science Laboratory Technology, Yaba College of Technology, Lagos, Nigeria

K. W. Fomba et al. (eds.), *Advances in Air Quality Research in Africa*,
Advances in Science, Technology & Innovation, https://doi.org/10.1007/978-3-031-53525-3_3

2 Materials and Methods

2.1 Study Area and Sampling Site

Ambient volatile organic compounds were collected in four locations namely; Westminster, Computer Village, Ojota, and Alaba market in the city. The air sampling method used for this work has been described elsewhere (Olumayede & Okuo, 2012). Briefly, a low volume sampling pump (Acuro, Drager, Lubeck, Germany) was used to draw air at an average flow rate of 0.51 min^{-1}. The tube was connected in front of the inlet of the adsorption tube. Samples were collected at the height of 1.5 m from the ground surface so as to reflect the human breathing zone. Air samples were collected at each site every four hours from 6.00 a.m. to 7.00 p.m.

2.2 Sample Preparation and Analysis

After sampling, halogenated hydrocarbons were desorbed using carbon disulphide. The extracted samples were analysed by GC-MS/FID system (GC, HP-7890A, Aligent Technology, California, USA). The VHHs were gasified and fed into the system to be separated using a DB-624 Chromatographic column (60 m × 0.25 mm i.d. × 1.8 μm film thickness). The temperature of the GC oven was initially set at 30 °C for 10 min, then increased to 120 °C at a rate of 5 °C/min. It was maintained at 120 °C for 7 min and again ramped to 200 °C at a rate of 5 °C/min for 16 min. Standard compounds and internal standards were used to build multi-point calibration curves for qualitative analysis of the VHHs. Quality control and quality assurances were based on detection limit, retention time, and analysis of reference standards.

3 Results and Discussion

3.1 Ambient Levels and Characteristics

Table 1 presents the profile of halogenated hydrocarbons detected and measured in the atmosphere of the four monitoring locations in Lagos. A total of 15 compounds were identified from the four monitoring locations in Lagos comprising four (4) chlorofluorocarbons, two (2) hydrochlorofluorocarbons, and six (6) chloroalkanes, 1 chloroalkene, two (2) halogenated benzenes.

The emission sources of these compounds had been previously attributed to the biotic activity of agricultural waste, along with the presence of oxidizing agents (Fe^{3+}) in such waste (Singh et al., 1977). 1,2-Dichloropropane was

Table 1 Concentration (mean ± SD) of halogenated hydrocarbons (μg/m^3) measured at different sites from Lagos, Nigeria

Categories/Compounds	Alaba international market	Ikeja computer village	Ojota dumpsite	Westminister market
Chlorofluorocarbon				
Trichloroflorо methane	1.36 ± 0.70	2.77 ± 0.92	1.01 ± 0.66	0.80 ± 0.54
Dichloro difluoro methane	3.25 ± 1.05	4.10 ± 1.05	2.4 ± 1.05	0.8 ± 1.05
Dichloro tetrafluoro ethane	2.18 ± 1.60	5.59 ± 1.12	1.60 ± 2.18	1.60 ± 2.18
1,1,2-Chlorotrifluoromethane	5.50 ± 2.04	2.33 ± 1.08	0.62 ± 0.16	nd
Hydrogenchlorofluoride				
Chlorodifluoromethane	0.1 ± 0.4	0.14 ± 0.10	0.15 ± 0.11	0.13 ± 0.11
1,1,1-Trichloroethane	0.05 ± 0.02	0.02 ± 0.01	0.94 ± 0.13	0.01 ± 0.01
Halogenated Alkane				
Dibromomethane	0.87 ± 1.44	1.70 ± 1.80	1.73 ± 1.25	1.03 ± 1.13
1,1,2-Trichloroethane	0.13 ± 0.34	0.02 ± 0.05	0.33 ± 0.15	0.02 ± 0.05
Tetrachloromethane	0.19 ± 0.13	0.18 ± 0.13	1.07 ± 0.03	0.06 ± 0.13
1,2-Dichloroethane	1.63 ± 1.09	1.76 ± 2.55	0.11 ± 0.60	0.91 ± 1.12
1,1,1-Trichloroethylene	0.74 ± 1.04	0.99 ± 1.55	1.51 ± 1.27	0.18 ± 1.83
1,2-Dichloropropane	2.08 ± 0.13	4.11 ± 0.13	0.17 ± 0.13	0.21 ± 0.13
Halogenated Alkene				
Chloroethene	0.43 ± 0.11	5.90 ± 1.42	3.68 ± 1.51	1.3 ± 0.98
Halogenated Benzene				
1,4-Dichlorobenzene	0.67 ± 0.15	0.98 ± 0.77	1.01 ± 0.83	0.64 ± 0.13
1,3-Dichlorobenzene	0.12 ± 0.03	0.04 ± 0.10	0.08 ± 0.04	nd

found in all the present study sites, with the highest mean value (4.11 μg/m^3) at the Ikeja Computer Village. This was closely followed by 2.08 μg/m^3 at Alaba International Market.

Surprisingly, a high concentration of tetrachloromethane, a natural organic chloride, was measured at Ojota dumpsite with an average mass concentration of 1.07 μg/m^3. Chlorinated benzenes are emitted during the combustion of chlorinated plastic PVC. The concentration of 1,4-dichloro benzene varied significantly among the sites, with the highest mean value of 1.01 μg/m^3 found at Ojota dumpsite, followed by 0.98 μg/m^3 at Ikeja Computer village and the lowest (0.64 μg/m^3) at Westminster International Market. The mass concentrations of the 15 VHHs observed in this study are higher than the reported global background levels (Qu-Yang et al., 2017).

3.2 Temporal Variation

The monthly average mass concentrations of VHHs in the ambient air are shown in Fig. 1. The average is seen to be higher at 3 locations in the month of June which is the peak of the rainy season in Nigeria. The poor diffusion conditions during this month might account for this observation (Zhang et al., 2017).

3.3 Diurnal Variation

Figure 2 clearly shows the diurnal variation of mass concentrations of total VHHs measured in this study. Chlorofluorocarbons show an increase from morning to afternoon period, then a decrease in the evening. The peak in the afternoon may be attributed to fresh emissions from dismantling, repairing activities, and combustion of wastes.

4 Conclusion

In this work, ambient volatile organic compounds were collected using an active sampling method with a low volume sampling pump from four different locations in Lagos, Nigeria between March and June 2019. The levels and variations of volatile halogenated hydrocarbons (VHHs) were assessed in the study. The results showed that 15 VHHs were detected in the atmosphere of Lagos city including four (4) chlorofluorocarbon, two (2) hydrochlorofluorocarbons, and six (6) chloroalkanes, 1 chloroalkene, two (2) Chlorinated benzenes. The monthly average concentration of the total VHHs ranged from 12.36 and 7.29 μg/m^3. Because of the serious environmental and human health impacts of halogenated hydrocarbon compounds, we suggest a long-term study of these pollutants in the area.

Fig. 1 Monthly average TVHH concentrations

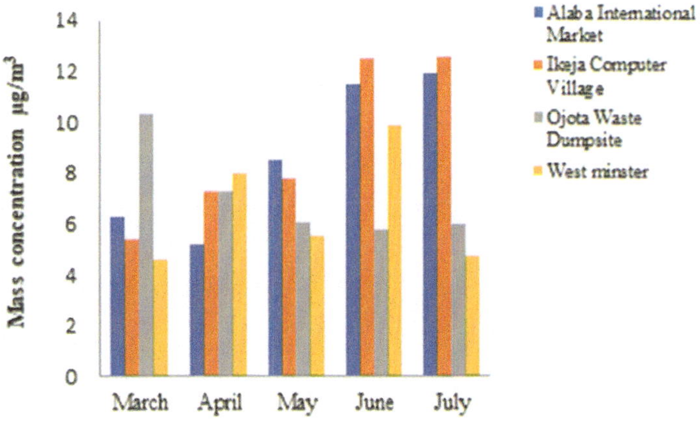

Fig. 2 Plot of diurnal variation of VHHs in Lagos

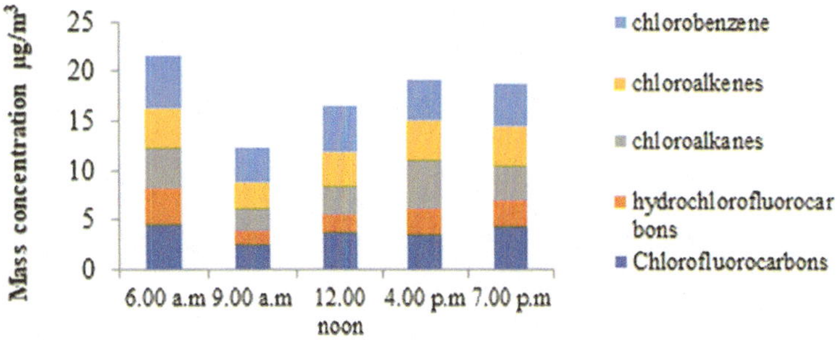

Acknowledgements The authors wish to specially thank the Department of Chemistry, Federal University Oye-Ekiti and Science Laboratory Department, Yaba College of Technology, Lagos for granting unhindered access to their facilities.

References

Lin, N., Kwarteng, L., Godwin, C., Warner, S., Robin, T., Arko-Mensah, Fobil, J. N., & Batterman, S. (2021). *Journal of Hazardous Materials, 419*, 126353.

Nnorom, I. C., & Osibanjo, O. (2008). *Waste Management, 28*, 1472–1479.

Ogungbuyi, O., Nnoron, I. C., Osinbanjo, O., & Schluep, M. (2012). E-waste country assessment Nigeria. In *E-waste Africa Project of the Secretariat of the Basel Convention*, Switzerland, p. 94.

Olumayede, E. G., & Okuo, J. M. (2012). Variation characteristics of volatile organic compounds in atmosphere of Benin City. *Polish Journal of Environmental Science, 21*(1), 77–186.

Qu-Yang, C., Chang, C., Wang, J., Shimada, K., Hatakeyama, S., Kato, S., Chiu, J., Sheu, G., & Lin, N. (2017). Characteristics of summertime volatile organic compounds in the lower free troposphere: Background measurements at Mt. Fuji. *Aerosol and Air quality Research, 17*, 3037–3051.

Singh, H. B., Salas, L. J., & Cavanagh, L. A. (1977). *Journal of Air Pollution Control Association, 27*(4), 332–336.

Zhang, G., Yao, B.,Vollmer, M. K., Montzka, S. A., Muhle, J., Weiss, R. F., O" Doherty, S., Fang, S., & Reimann, S. (2017). Ambient mixing ratios of atmospheric halogenated compounds at five background stations in China. *Atmospheric Environment, 160*, 55–69.

Impact of Mineral Dust Pollution on the Environmental Nutrient Budget in Cabo Verde

Nongma Kaboré, Khanneh Wadinga Fomba, Corrine Almeida, Daniel Tetteh Quaye and Hartmut Herrmann

Abstract

Mineral dust is a source of pollution and air quality. However, it can be a source of nutrients to the terrestrial and marine ecosystems. In this study, the impact of dust pollution on some essential nutrients has been investigated. Particulate Matter (PM_{10}) samples were collected over twelve days at Cape Verde Atmospheric Observatory, from 27 January to 8 February 2022, using a low-volume sampler on 37 mm diameter quartz fiber filters. The concentration of water-soluble inorganic ions in PM_{10} was determined using ion chromatography. The following inorganic nutrients, Fe, PO_4^{2-}, NO_3^-, and NH_4^+, were quantified due to their high influence on the marine ecosystem. The concentration of PM_{10} drastically increased during the Saharan dust intrusion. The mean value during the Saharan dust days was 118.9 µg/m^3 (for n = 6 days during the dust storm) compared to the background value of 2.04 µg/m^3 (for n = 3 days before the dust storm). The daily concentrations of PO_4^{2-}, NO_3^-, and NH_4^+ in dust days were factor 2 higher than those of non-dust days. These results show a strong effect of dust pollution on the ecosystem that needs further attention.

Keywords

Saharan dust storm · PM_{10} · Air pollution · Inorganic nutrients

N. Kaboré · C. Almeida · D. T. Quaye
West African Science Service Centre on Climate Change and Adapted Land Use (WASCAL)-Universidade Técnica do Atlântico (UTA), Campus de Ribeiro de Julião, CP 163 Mindelo, São Vicente, Cabo Verde

K. W. Fomba · H. Herrmann (✉)
Atmospheric Chemistry Department (ACD), Leibniz Institute for Tropospheric Research (TROPOS), Permoserstraße 15, 04318 Leipzig, Germany
e-mail: herrmann@tropos.de

1 Introduction

The atmospheric nutrient deposition encompasses macronutrients such as phosphate, nitrogen, and micronutrients like iron. These nutrients regulate biological, chemical, and geological marine processes. The atmospheric input is significant in oligotrophic ocean regions like the north tropical Atlantic. The nitrogen inputs from both atmosphere and rivers are equivalent (Duce et al., 1991). In contrast, atmospheric phosphorus deposition is significant, especially in the coastal region (Zamora et al., 2013). Human activities and natural climate change have contributed to an increase in the emission of particles from anthropogenic and natural activities, increasing the amount of atmospheric deposition of nutrients into oceans (Xing et al., 2017).

Despite a large amount of literature available on the topic of fluxes of atmospheric inorganic nutrients and their impact on the marine ecosystem for an extended period, we still lack enough knowledge about the local variation of PM_{10} inorganic nutrients over dust outbreak periods in different systems in Africa, especially in Cabo Verde.

The fundamental goal of this study was to analyze PM_{10} atmospheric inorganic ionic species fluxes before, during, and after a short-time dust outbreak.

2 Material and Methods

2.1 Site and Sampling Description

Sampling of atmospheric aerosols was conducted from 27 January to 8 February 2022 at the Cape Verde Atmospheric Observatory (CVAO) (16° 51′ 49" N, 24° 52′ 02" W) situated in Sao Vicente Island in Cape Verde. (Fomba et al., 2014) Have described in detail the sampling site.

The sampling period was marked by six days of dust storms from 31 January to 5 February, dividing the sampling period into two dust-free days: before and after the

K. W. Fomba et al. (eds.), *Advances in Air Quality Research in Africa*,
Advances in Science, Technology & Innovation, https://doi.org/10.1007/978-3-031-53525-3_4

dust storms from 27 to 29 January and 6 to 8 February 2022, respectively. PM_{10} samples were collected for 24 h using a PM_{10} impactor inlet (PEM) and a pump sampling at 10 L min^{-1}, mounted at the top of a 30 m tower. Samples were collected on 37 mm diameter quartz fiber filters (Whatman, quartz filter). After sampling, the filters were transported and kept in a −20 °C freezer. The samples were transported from Cabo Verde to Leipzig (Germany) under freezing conditions (at −20 °C) for chemical analysis.

2.2 Chemical Analysis of Water-Soluble Inorganic Ions

Two spots of 10 mm forming a 1.57 cm^2 area from each filter have been punched out of the samples. The spots were placed into labeled vials, and 2 ml of ultrapure deionized water was added. The samples were shaken for 120 min. The solutions were filtered through a 0.45 µm syringe filter to eliminate colloidal particles that could eventually block the chromatographic columns.

An ion chromatographer (DIONEX, Model ICS 6000, USA) was used to analyze the concentration of significant anions (SO_4^{2-}, NO_3^-, NO_2^-, PO_4^{3-}, Cl^-, F^-) and cations (NH_4^+, Na^+, Mg^{2+}, Ca^{2+}, K^+).

3 Results and Discussion

3.1 Concentration of Water-Soluble Ion Species in PM_{10}

Table 1 presents the mean concentrations of the water-soluble inorganic ions based on the twelve samples. From the data, it was seen clearly that the mean ionic concentrations in PM_{10} followed the order $Na^+ > SO_4^{2-} > Cl^- > NO_3^- > Ca^{2+} > K^+ > F^- > Mg^{2+} > NH_4^+ > PO_4^{3-} > NO_2^-$. The results showed a high increase in concentration of sea salt due to the proximity of the study area to the sea. The PM_{10} mass concentration showed strong variations over time with a concentration in the range of 0.7 to 241.3 µg/m^3 with the highest possible standard deviation from the mean value of 68.2 ± 69.1 µg/m^3, which was above the limits stated in the World Health Organization (WHO) air quality guidelines (45 µg/m^3).

The PM_{10} concentration time series continuously shows six days of dust outbreaks (PM_{10} plot not shown here) from 31 January to 5 February. The PM_{10} maximum value of 241.3 µg/m^3 is between the maximum values of 227 µg/m^3 and 250 µg/m^3 observed by Niedermeier et al. (2014), and (Ginoux et al., 2001) models, during different dust events, respectively. The results were also higher than the 1-year mean (48 ± 64 µg/m^3) reported by Pio et al. (2014), who conducted a 1-year (2011) continuous field measurement

Table 1 Summary of the PM_{10} aerosol ionic species mass concentrations during the sampling period. The minimum (Min), maximum (Max), average (Mean), and standard deviation (Std) values of the species are in ng/m^3, except for PM_{10}, Cl^-, NO_3^-, SO_4^-, Na^+, and Fe, where the values are in µg/m^3

	Min	Max	Mean	Std
PM_{10}	0.7	241.3	68.2	69.1
Cl^-	1.5	9.1	4.5	2.3
NO_3^-	0.3	2.4	1.3	0.7
SO_4^{2-}	1.9	9.4	4.7	1.9
NO_2^-	4.9	21.6	13.8	4.4
F^-	122.7	284.3	216.4	47.4
PO_4^{3-}	15.8	64.4	47.5	14.0
Na^+	6.9	17.7	13.9	2.7
NH_4^+	28.6	92.8	65.4	17.9
K^+	67.5	475.5	255.5	122.0
Mg^{2+}	1.8	290.7	76.5	90.7
Ca^{2+}	12.7	2749.4	738.8	807.3
Fe	0.04	5.25	1.68	1.98

campaign on Santiago Island, Cabo Verde, as part of the CV-DUST implemented project.

3.2 Influence of the Saharan Dust Storm on the Key Inorganic Species

The Saharan dust storm is a factor contributing to the Atlantic Ocean ecosystem. As mentioned in Sect. 2.1, dust events occurred during the sampling period, which amounted to 6 days. Figure 1 shows the daily concentrations of dissolved inorganic N and P species during both Saharan (shaded rectangles) and non-Saharan dust periods. Previous studies indicated that the concentration of ionic species in PM_{10} dramatically increased when dust phenomena were observed over CVAO (Fomba et al., 2014).

Our result also suggested that dissolved inorganic N and P in PM_{10} were marked by a high increase in the Saharan dust period. Note that the mean concentrations of the dusty period (for $n = 6$ days during the dust storm) of NO_3^-, PO_4^{3-}, and NH_4^+ were 1.71, 0.06, and 0.07 µg/m^3, respectively. These values were almost twice the corresponding values of 0.32, 0.03, and 0.04 µg/m^3, respectively during dust-free days (referred to the $n = 3$ days before the dust outbreaks). This indicates that the Saharan dust intrusion was responsible for a significant increase in the dissolved inorganic nutrient levels at the monitoring site of this study. The dust had a pronounced positive impact on Fe, with a mean of 3.04 µg/m^3 during the dusty period compared to the referred days before dust event (0.11 µg/m^3), suggesting that the dust storm was a primary contributor of Fe.

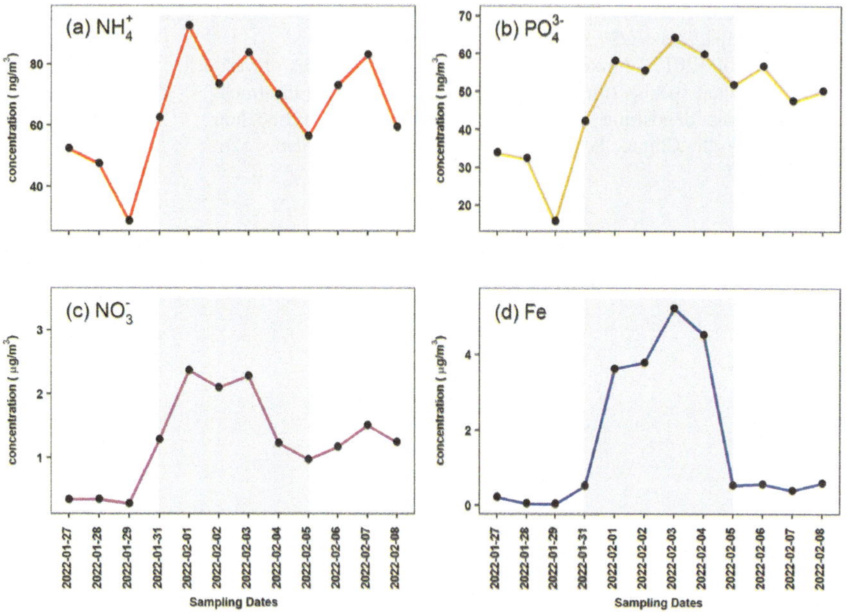

Fig. 1 Variations of the dissolved inorganic N and P species concentrations in PM_{10} during the sampling period. Figures (**a**) and (**b**) for ammonium and phosphate are in ng/m^3, and (**c**) and (**d**) for nitrate and iron, respectively, are in µg/m^3. The shaded area is the dust storm days

We also found a slow deposition of species such as PO_4^{3-} and NH4$^+$, i.e., those species seemed to remain in the atmosphere after the dust intrusion, while the depositions of the NO_3^- and Fe were relatively quick. Our results were consistent with many studies that reported significantly increased mass concentrations of Fe, NO_3^-, and NH_4^+. The increase is attributable to the long-distance transport of dust particles from emission sources or to physical mixing and the heterogeneous formation on the dust surfaces (Liu & Bei, 2016; Qi et al., 2018; Wang et al., 2018). Several previous studies reported a significant increase in the concentration of Fe during dust episodes (Liu & Bei, 2016; Qi et al., 2018). Similarly, in our study, Fe also increased during dust episodes, indicating the influence of dust pollution.

4 Conclusion

Investigation of PM_{10} filter samples collected from 27 January to 8 February 2022 provides information on dust pollution's impact on the environmental nutrient budget over Cabo Verde. The Saharan outflow distinctly influenced nutrient fluxes to the ocean surrounding Cabo Verde. PM_{10} concentrations were elevated during days of the dust storm, but the elevation was more marked on 1st of February, with the daily PM_{10} value reaching 241.3 µg/m^3. The change of inorganic ion concentrations in PM_{10} aerosol was mainly affected by dust, especially with Fe increased by 91% during the dust storm.

Acknowledgements We are very grateful to Leibniz Institute for Tropospheric Research, Germany and the Cape Verde Atmospheric Observatory (CVAO) for their help with the atmospheric data and laboratory resources. We express gratitude to the Atlantic Technical University and Wascal Cape Verde staff for their support and advice.

References

Duce, R. A., et al. (1991). The atmospheric input of trace species to the world ocean. *Global Biogeochemical Cycles, 5*(3), 193–259.

Fomba, K. W., Müller, K., Van Pinxteren, D., Poulain, L., Van Pinxteren, M., & Herrmann, H. (2014). Long-term chemical characterization of tropical and marine aerosols at the Cape Verde atmospheric observatory (CVAO) from 2007 to 2011. *Atmospheric Chemistry and Physics, 14*(17), 8883–8904. https://doi.org/10.5194/acp-14-8883-2014

Ginoux, P., Chin, M., Tegen, I., Goddard, T., Holben, B., Dubovik, O., & Lin, S. -J. (2001). Sources and distributions of dust aerosols simulated with the GOCART model. In this model all topographic from the Goddard Earth Observing System Data Assimilation System (GEOS emission is estimated to be between. *Journal of Geophysical Research: Atmospheres, 106*, 20255–20273.

Liu, Q., & Bei, Y. (2016). Impacts of crystal metal on secondary aliphatic amine aerosol formation during dust storm episodes in Beijing. *Atmospheric Environment, 128*, 227–234. https://doi.org/10.1016/j.atmosenv.2016.01.013

Niedermeier, N., et al. (2014). Mass deposition fluxes of Saharan mineral dust to the tropical northeast Atlantic Ocean: An intercomparison of methods. *Atmospheric Chemistry and Physics, 14*(5), 2245–2266. https://doi.org/10.5194/acp-14-2245-2014

Pio, C. A., et al. (2014). Seasonal variability of aerosol concentration and size distribution in Cape Verde using a continuous aerosol optical spectrometer. *Frontiers in Environmental Science, 2*(May), 1–11. https://doi.org/10.3389/fenvs.2014.00015

Qi, J., et al. (2018). The concentration, source and deposition flux of ammonium and nitrate in atmospheric particles during dust events at a coastal site in northern China. *Atmospheric Chemistry and Physics, 18*(2), 571–586. https://doi.org/10.5194/acp-18-571-2018

Wang, Q., et al. (2018). Environmentally dependent dust chemistry of a super Asian dust storm in March 2010: Observation and simulation. *Atmospheric Chemistry and Physics, 18*(5), 3505–3521. https://doi.org/10.5194/acp-18-3505-2018

Xing, J. et al. (2017). Fluxes, seasonal patterns and sources of various nutrient species (nitrogen, phosphorus and silicon) in atmospheric wet deposition and their ecological effects on Jiaozhou Bay, North China. *Science of the Total Environment, 576*, 617–627.

Zamora, L. M., Prospero, J. M., Hansell, D. A., & Trapp, J. M. (2013). Atmospheric P deposition to the subtropical North Atlantic: Sources, properties, and relationship to N deposition. *Journal of Geophysical Research Atmospheres, 118*(3), 1546–1562. https://doi.org/10.1002/jgrd.50187

Impact of Urban Emissions on Regional Air Quality in Fez City Area, Morocco

Nabil Deabji, Khanneh Wadinga Fomba, Laurent Poulain, Eduardo José dos Santos Souza, Abdelwahid Mellouki and Hartmut Herrmann

Abstract

An intensive field campaign was conducted to investigate the chemical composition of particles simultaneously at the remote Mohammed V observatory (AMV) and the urban city of Fez in the Middle Atlas region of Morocco. This study aimed to assess the regional air quality and health impact of exposure to PM_{10}. Aerosol sampling was carried out using a particle collector PM_{10}. Various laboratory analyses were performed, including particle mass, OC/EC, inorganic ions, trace metals, and a wide range of organic species. The results show that the PM mass has a similar trend at both sites. However, the mass concentration was twice as high at the Fez urban site (62 μg m$-^{3}$) than at AMV (35 μg m^{-3}). Anthropogenic metals, dominated by Cr, Cu, and Ni, increased by about 62% at the urban site of Fez. PAHs at Fez were two times higher with high contributions of Fluorene, Retene, Phenanthrene, and Benz (a) anthracene, resulting mainly from vehicular emission, typical for industrial emission, waste incineration, and other combustion processes.

Keywords

Air pollution · Anthropogenic metals · PAHs

1 Introduction

The World Health Organization has estimated that ambient air pollution is responsible for seven million premature deaths per year worldwide (World Health Organization, 2021). Some field studies in African cities have reported annual average PM_{10} concentrations that exceed permissible levels (Naidja et al., 2018). However, there is still a real need for air quality monitoring facilities to assess the risk to human health to improve air quality and control emissions. Most particulate matter in African urban areas comes from road traffic, biomass burning of municipal waste, and heavy industry, such as cement plants (Naidja et al., 2018). Previous studies in Morocco showed a high contribution from local biomass combustion rather than long-distance transport (Benchrif et al., 2018). Another study reported that ozone concentrations reached alarming levels, and PM_{10} levels exceeded threshold limits during the summer and spring (Inchaouh et al., 2017). However, in most cases, air quality is only quantified using a combination of monitored and modeled data without considering the health impact assessment. Thus, studies investigating the chemical composition of particles are needed in Morocco. Consequently, the acquisition of new knowledge and reduction of uncertainty regarding the effect of aerosols on health and atmospheric composition related to air pollution. This study aims to collect information on PM levels and their chemical composition. It also aims to identify the different sources of PM emissions that contribute to the deterioration of air quality in Morocco and their health impact, which will allow the development of abatement strategies to improve local air quality.

N. Deabji · K. W. Fomba · L. Poulain · E. J. dos Santos Souza · H. Herrmann (✉)
Atmospheric Chemistry Department (ACD), Leibniz Institute for Tropospheric Research (TROPOS), Permoserstraße 15, 04318 Leipzig, Germany
e-mail: herrmann@tropos.de

N. Deabji
e-mail: deabji@tropos.de

A. Mellouki
University Mohamed 6 Polytechnic, Lot 660, Hay Moulay Rachid, Ben Guerir, Morocco

Institut de Combustion Aérothermique Réactivité Et Environnement, OSUC-CNRS, 1C Avenue de La Recherche Scientifique, 45071 Orléans CEDEX 2, France

K. W. Fomba et al. (eds.), *Advances in Air Quality Research in Africa*,
Advances in Science, Technology & Innovation, https://doi.org/10.1007/978-3-031-53525-3_5

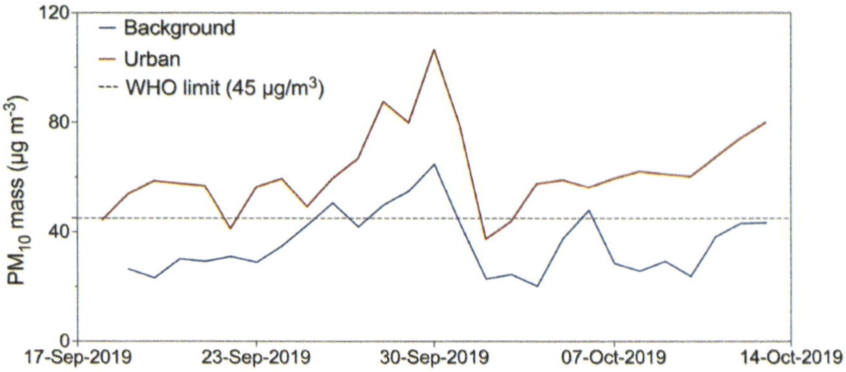

Fig. 1 Temporal variation of PM_{10} levels at the background and urban sites in Morocco

2 Materials and Methods

Aerosol sampling was carried out simultaneously during an intensive campaign at two sites in the Middle Atlas region of Morocco: The Atlas Mohammed V Observatory (AMV), a newly established research station located at a high altitude (2100 m.a.s.l) in the Atlas Mountains, and an urban site located in the city center of Fez. Particulate matter (PM_{10}) filter samples were collected at both sites using a high-volume (HV) collector during an intensive measurement campaign from the 15th September to 15th October 2019. The chemical composition of the samples was analyzed for PM mass, trace metals, water-soluble ions, organic carbon (OC/EC), and polycyclic aromatic hydrocarbon (PAH) contents. Comprehensive details of experimental procedures employed in this study follow the methodology outlined elsewhere in the literature (Deabji et al., 2021).

3 Results and Discussion

3.1 PM_{10} Mass Variation

The daily average amount of PM_{10} at the study site was 62 µg m^{-3}, ranging from 37.5 µg m^{-3} (02/10/2019) to 107 µg m^{-3} (30/09/2019) (see Fig. 1).

The average PM_{10} concentration at Fez was twice as high as the regional background concentration (35 µg m^{-3}), and about 81% of the samples exceeded the World Health Organization (WHO) guideline values which recommend a concentration limit of 45 µg m^{-3} in 24 h (WHO, 2021). The average PM_{10} concentration found in the city of Fez is comparable with studies done at other Moroccan sites, such as Marrakech, and Kenitra, which are exposed to high urban emissions, with PM_{10} concentrations between 50 and 110 µg m^{-3} (Inchaouh et al., 2017; Tahri et al., 2013).

3.2 Chemical Composition

The chemical composition of PM_{10} including OC/EC and major inorganic elements is presented in the following Table 1.

The concentration of OC recorded at the remote site of AMV is 1.6 µg m^{-3}, which is mainly due to biogenic emissions. Moreover, the OC concentration was 5 times higher in Fez (7.2 µg m^{-3}) indicating the strong contribution of other potentially anthropogenic sources to biogenic emissions. The low concentration of EC at AMV (0.29 µg m^{-3}) indicates that the local anthropogenic influence is minimal, whereas the average EC concentration recorded at Fez (2.6 µg m^{-3}) was 10 times higher. The average concentration of sulfate and nitrate recorded at AMV which is about 1.1 µg m^{-3} was 2 times higher than that at Fez. Comparable ammonium concentrations were found at both sites with a higher concentration at AMV during specific periods (Deabji et al., 2021).

Table 1 Table captions should be placed above the tables

Chemical component	Background (AMV)	Urban (Fez City)
Mass	35	62
OC	1.6	7.2
EC	0.29	2.6
Cl−	0.06	0.44
NO_3^-	1.1	2.9
SO_4^{2-}	1.1	2.5
$C_2O_4^{2-}$	0.1	0.18
Na^+	0.19	0.77
NH_4^+	0.40	0.71
K^+	0.06	0.55
Mg^{2+}	0.06	0.15
Ca^{2+}	0.81	2.23

Fig. 2 Variation of anthropogenic metals and PAHs at background and urban sites

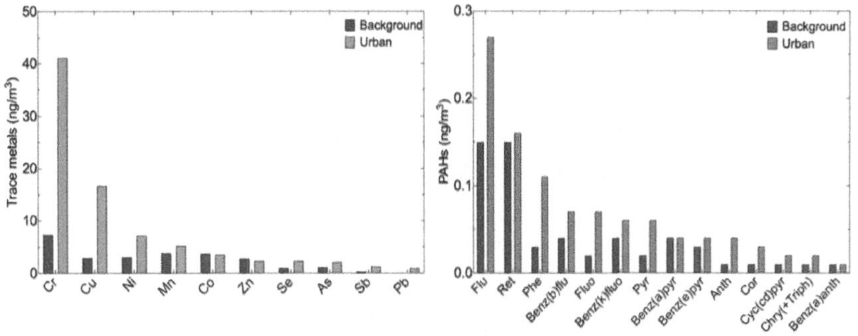

3.3 Impact of Air Pollution

Levels of some trace metals and PAHs at Fez were significantly higher than WHO limit values, as indicated in Fig. 2. For instance, The AMV site was used as a reference site to highlight the impact of air pollution at Fez. Therefore, Cr, Cu, and Ni values were 2–5 times higher than those found at AMV. While PAHs which are dominated by Fluorene, Retene, and Phenanthrene are on average 2–4 times higher. This indicates that the city of Fez is exposed to high levels of pollution most likely from waste incineration and traffic emissions. Health risk assessments were performed using health index such as Hazards Quotient (HQ) and carcinogenic (ELCR) index for both receptors (adult and child). Description of health index. The calculated values of Hazards Quotient (HQ) were found higher for all metals especially Ni, Mn, and Cr which have adverse health effects. It was observed that high traffic intensity may have been responsible for the high contribution of trace metals. The highest value of ELCR was found for lead, the trend of the average value of carcinogenic risk factor was found to be ~2–4 times higher than the prescribed limit.

4 Conclusion

The urban city of Fez is exposed to high levels of particulate pollution from combustion processes and vehicular emissions, which can have a drastic impact on the health of the population. These findings are of interest to managing air quality, as they highlight the critical emission sources on which local authorities should focus their abatement efforts.

Acknowledgements The authors would like to thank all contributors for their efforts in sample collection, the TROPOS laboratory assistants for their helpful work, and Horizon 2020 (MARSU) for their financial support.

References

Benchrif, A., Guinot, B., Bounakhla, M., Cachier, H., Damnati, B., & Baghdad, B. (2018). Aerosols in Northern Morocco: Input pathways and their chemical fingerprint. *Atmospheric Environment, 174* (February), 140–47. Research, 25(13), 12303–12328.

Deabji, N., Fomba, K. W., El Hajjaji, S., Mellouki, A., Poulain, L., Zeppenfeld, S., & Herrmann, H. (2021). First insights into northern Africa high-altitude background aerosol chemical composition and source influences. *Atmospheric Chemistry and Physics, 21*(24), 18147–18174.

Inchaouh, M., Tahiri, M., El Johra, B., & Abboubi, R. (2017). State of ambient air quality in Marrakech city (Morocco) over the period 2009–2012. *GEOMATE Journal, 12*(29), 99–106.

Naidja, L., Ali-Khodja, H., & Khardi, S. (2018). Sources and levels of particulate matter in North African and Sub-Saharan cities: A literature review. *Environmental Science and Pollution*.

Tahri, M., Bounakhla, M., Zghaïd, M., Noack, Y., Benyaïch, F., & Benchrif, A. (2013). Evaluation of airborne particulate matter pollution in Kenitra City Morocco. *Revista Ambiente & Água, 8*(1), 38–47.

World Health Organization. (2021). WHO global air quality guidelines: particulate matter ($PM_{2.5}$ and PM_{10}), ozone, nitrogen dioxide, sulfur dioxide and carbon monoxide. World Health Organization.

Aerosols Optical Properties Profile Characterization Over São Vicente, Cabo Verde, During the ASKOS Campaign

Samira Moussa Idrissa, Nilton Évora do Rosário and Nikolaos Siomos

Abstract

Studying the vertical distribution of aerosol particles is an important aspect for understanding their influence on the climate system. One way to evaluate the impact of aerosols is to discriminate their source and optical properties. The main objective of this study is to characterize and analyze aerosol's optical properties profile under distinct aerosol mixture states over São Vicente Island, Cabo Verde, during the ASKOS campaign that took part from July to September of 2021. A recently developed method, the Elastic Extinction (ElEx), which allows aerosol backscattering profile separations by type and estimation of total extinction coefficient, has been applied. Conducting a case study of pure Saharan dust dominance in the free troposphere and marine particles in the Marine Boundary Layer (MBL), the ElEx method revealed the ability to discriminate consistently the vertical distribution of the two types of aerosols.

Keywords

Aerosol types · Depolarization ratio · Lidar ratio · Saharan dust

1 Introduction

Aerosols are an essential component of the climate system (Knippertz & Todd, 2012). They have the capacity to influence climate in various ways. Due to their optical radiative properties and their potential to act as cloud condensation nuclei (CCN) and ice nucleating (IN) particles, they play a central role in Earth's energy budget and hydrological cycle (Ramanathan et al., 2007). Additionally, as a component of air pollution, aerosol particles pose a serious risk to human health, mainly in regions close to their source's emission and downwind areas. Considering their climatic and environmental implications, the monitoring and modeling of atmospheric aerosols are of fundamental importance. However, emission sources diversity, transport, and atmospheric mixing of aerosol particles worldwide pose an extremely challenging task to the existent sparse monitoring network, which brings great levels of uncertainty for their characterization and modeling. To reduce these uncertainties, coordinated and strategic integration of data from multiple platforms (e.g., ground-based, satellite, ship, and aircraft) (Holben et al., 1998) and techniques (e.g., in-situ measurement, remote sensing, numerical modeling, and data assimilation) are highly required. Within this perspective, the present study aimed to study the vertical distribution of aerosol mixture and optical properties in the atmosphere column over São Vicente Island, Cabo Verde, by using the state-of-the-art polarization lidar system named eVe (Paschou et al., 2022) and by applying the method called ElEx (Elastic Extinction) recently developed (Giannakaki et al., 2020).

S. M. Idrissa
Atlantic Technical University, Mindelo, São Vicente, Cabo Verde
e-mail: sidrissa@uta.cv

N. É. do Rosário (✉)
Environmental Sciences Department, Sao Paulo Federal University, Diadema, Brazil
e-mail: nrosario@unifesp.br

N. Siomos
Meteorological Institute, Ludwig-Maximilians-Universität, Munich, Germany
e-mail: nikolaos.siomos@lmu.de

K. W. Fomba et al. (eds.), *Advances in Air Quality Research in Africa*,
Advances in Science, Technology & Innovation, https://doi.org/10.1007/978-3-031-53525-3_6

2 Data and Methodology

2.1 Data Collection

The dataset considered in this investigation was collected at Mindelo City, São Vicente Island, Cabo Verde, during the JATAC/ASKOS experiment organized by the European Space Agency (ESA) during the West African Monsoon period of 2021(from July to September). The data were obtained using the eVe Raman lidar deployed by the National Observatory of Athens (NOA) in Mindelo.

2.2 Data Analysis

To obtain the aerosol vertical profile discrimination we used the EIEx method recently developed (Giannakaki et al., 2020). This method allows the vertical discrimination of two types of aerosols and for that requires information of their individual depolarization and lidar ratios. These values should be fixed and known from the literature. Three steps are needed to perform EIEx (Giannakaki et al., 2020):

First Step: Backscatter coefficient and particle depolarization

To decompose the total retrieved backscatter profile of eVe into two aerosol components by using the equation of the backscatter (Eq. 1) and applying the corresponding depolarization ratio of each aerosol component chosen from the literature (Table 1).

Second step: Estimation of extinction coefficient profile

The separation of the backscatter profile by components gives us the opportunity to calculate the total extinction coefficient by applying the proper lidar ratio of each aerosol component (Eq. 2).

$$\beta_1 = \frac{\beta_t(\delta_t - \delta_2)(1 + \delta_1)}{(\delta_1 - \delta_2)(1 + \delta_t)} \quad (1)$$

where β_t, β_1 and β_2 represent, respectively, the backscatter coefficient profiles of the total aerosol mixture, the first and the second aerosol types. δ_t, δ_1 and δ_2 are the particle depolarization ratio of the mixture of aerosols, the first and the second aerosol types, respectively.

$$a_t = a_1 + a_2 = \beta_1 \cdot lr_1 + \beta_2 \quad (2)$$

where a_t, a_1 and a_2 represent, respectively, the extinction coefficient profiles of the total aerosol mixture, of the first and the second aerosol types. lr_1 is the lidar ratio of the first aerosol type.

Third Step: EIEx validation

Finally, the EIEx method results are validated by comparing the profiles obtained with those from the Raman inversion method using eVe measurements.

3 Results and Discussion

The air mass trajectories analysis indicates that the air masses reaching Mindelo in the marine boundary layer have traveled all their way close to the ground from the Canary Island during the previous three days (Fig. 1). Therefore, particles in this layer are expected to be mainly from the MBL. In contrast, the air masses that arrived over Mindelo in the layers above MBL (1000, 2000, 3000, and 4000 mts) came from the Sahara Desert, and hence carried dust particles.

Figure 2 presents the profile of optical parameters obtained using the Raman method and the results of profiles separation by aerosol types using the EIEx method. The lidar ratio profile for 355 nm obtained by applying the Raman method indicates two aerosol layers: the first mostly confined to the MBL (<1 km), and the second above MBL in layer from 1 to 4 km. The eVe profile indicates values of lidar ratio close to 53 sr from ~1 to 4 km height, which is consistent with dust presence. Below about 1 km, the lidar ratio declines to around 18 sr (355 nm) inside the MBL. This reduction of the lidar ratio values from 53 to 18 sr points out marine aerosols as the major non-dust component in the MBL.

The lidar ratio and extinction coefficient obtained from the EIEx retrieval showed a better agreement with eVe Raman inversions for the mixture of dust and marine particles than for the mixture of dust and pollution particles.

Table 1 Typical values of lidar and depolarization ratios for distinct aerosol types taken from the literature

Aerosol types	Lidar ratio (lr)	Depolarization ratio (δ)	References
Marine	18 ± 04 sr	0.02 ± 0.010	Groß et al. (2011)
Sahara dust	53 ± 10 sr	0.26 ± 0.06	Tesche et al. (2011)
Pollution	44 ± 05 sr	0.015 ± 0.01	Kaduk (2015)

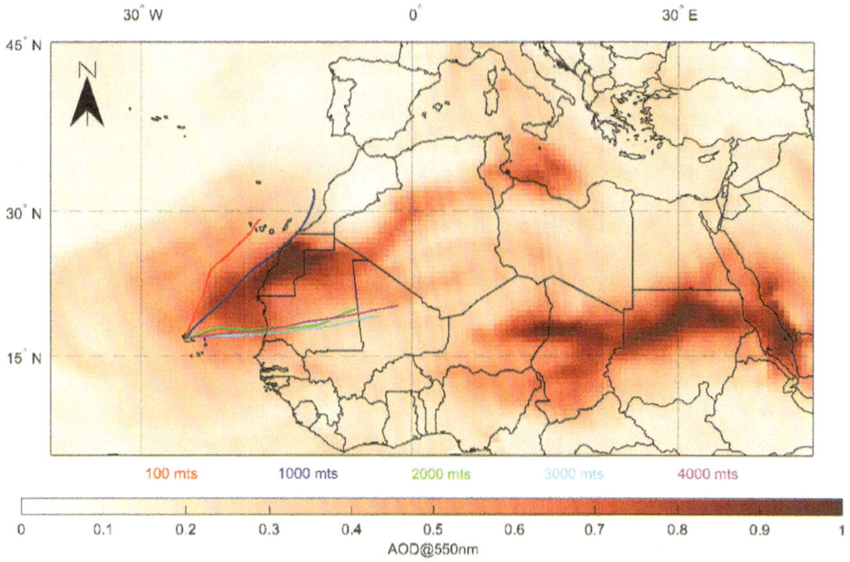

Fig. 1 Backward trajectories of air masses arriving over São Vicente at different altitudes on the day 10 September 2021 on top of mean aerosol optical depth at 550 nm (AOD@550 nm) map from MERRA-2 reanalyzes for the same day

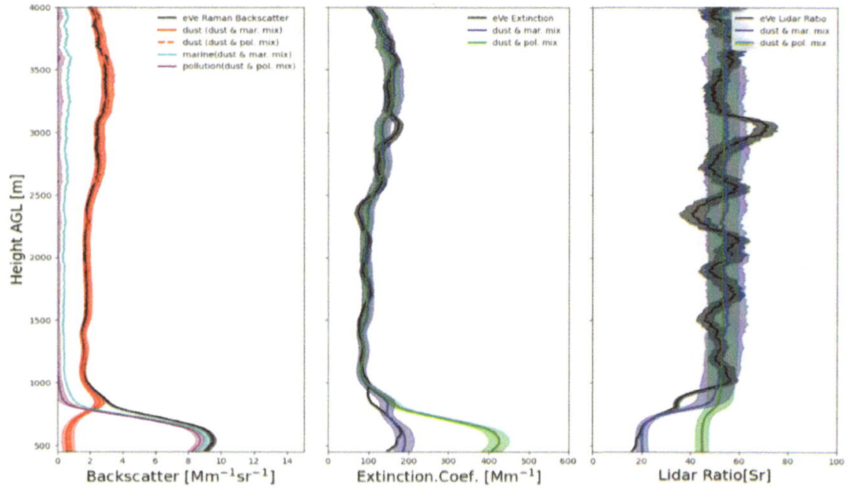

Fig. 2 Reference lidar ratio and extinction and backscatter coefficients profiles at 355 nm from eVe (black line) for the case of 10 September 2021 and resolved profiles using dust (red), marine (cyan), and pollution (purple) aerosol types and considering different assumptions for aerosol mixtures applying EIEx method: dust/marine (blue) and dust/pollution (green)

However, some mismatches are noticeable in the transition layer between the layer dominated by marine particles (MBL) and layers above dominated by dust particles. This may be caused by the presence of smoke or even pollution aerosols.

4 Conclusion

The composition of vertical distribution of aerosol particles is essential for understanding their influence on the climate system. Cabo Verde islands are in a unique position to study the mixture and interaction of dust, marine, and pollution aerosols in the atmosphere column. This study sought to analyze the vertical distribution of aerosol optical properties as a function of aerosol types over São Vicente Island, in Cabo Verde, using the EIEx separation method applied to data from the ASKOS campaign. The comparison of the derived EIEx lidar ratio and extinction coefficient profiles discriminated by aerosol types with the reference Raman inversions from eVe showed that EIEx method provides consistent results.

References

Giannakaki, E., Kokkalis, P., Marinou, E., Bartsotas, N. S., Amiridis, V., Ansmann, A., & Komppula, M. (2020). The potential of elastic and polarization lidars to retrieve extinction profiles. *Atmospheric Measurement Techniques*, *13*(2), 893–905. https://doi.org/10.5194/AMT-13-893-2020

Groß, S., Gasteiger, J., Freudenthaler, V., Wiegner, M., Geiß, A., Schladitz, A., Toledano, C., Kandler, K., Tesche, M., Ansmann, A., & Wiedensohler, A. (2011). Characterization of the planetary boundary layer during SAMUM-2 by means of lidar

measurements. *Tellus b: Chemical and Physical Meteorology, 63*(4), 695–705. https://doi.org/10.1111/j.1600-0889.2011.00557.x
Holben, B. N., Eck, T. F., Slutsker, I., Tanré, D., Buis, J. P., Setzer, A., Vermote, E., Reagan, J. A., Kaufman, Y. J., Nakajima, T., Lavenu, F., Jankowiak, I., & Smirnov, A. (1998). AERONET—A federated instrument network and data archive for aerosol characterization. *Remote Sensing of Environment, 66*(1), 1–16. https://doi.org/10.1016/S0034-4257(98)00031-5
Kaduk, C. (2017). Characterization of the optical properties of complex aerosol mixtures observed with a multiwavelength–Raman–polarization lidar during the 6–weeks BACCHUS campaign in Cyprus in spring 2015.
Knippertz, P., & Todd, M. C. (2012). Mineral dust aerosols over the Sahara: Meteorological controls on emission and transport and implications for modeling. *Reviews of Geophysics, 50*(1). https://doi.org/10.1029/2011RG000362
Paschou, P., Siomos, N., Tsekeri, A., Louridas, A., Georgoussis, G., Freudenthaler, V., Binietoglou, I., Tsaknakis, G., Tavernarakis, A., Evangelatos, C., Von Bismarck, J., Kanitz, T., Meleti, C., Marinou, E., & Amiridis, V. (2022). The eVe reference polarisation lidar system for the calibration and validation of the Aeolus L2A product. *Atmospheric Measurement Techniques, 15*(7), 2299–2323. https://doi.org/10.5194/AMT-15-2299-2022
Ramanathan, V., Ramana, M. V., Roberts, G., Kim, D., Corrigan, C., Chung, C., & Winker, D. (2007). Warming trends in Asia amplified by brown cloud solar absorption. *Nature 2007 448:7153, 448*(7153), 575–578. https://doi.org/10.1038/nature06019
Tesche, M., Müller, D., Gross, S., Ansmann, A., Althausen, D., Freudenthaler, V., Weinzierl, B., Veira, A., & Petzold, A. (2011). Optical and microphysical properties of smoke over Cape Verde inferred from multiwavelength lidar measurements. *Taylor & Francis, 63*(4), 677–694. https://doi.org/10.1111/j.1600-0889.2011.00549.x

Ambient and Indoor Air Quality Measurements

Real-Time Hospital Air Quality Monitoring Based on a Low-Cost Smart Electronic Device and Health Risk Assessment

Jacob Mbarndouka Taamté, Nasser Nducol, Gouroudja Ahmadou, René Claude Etende Essama, Michaux Kountchou Noubé, Yvette Flore Tchuente Siaka and Saïdou

Abstract

This paper reports the real-time air quality measurements and health risk assessment of exposed people in two reference hospitals of Yaoundé City (Biyem-assi District Hospital and Yaoundé Central Hospital), by using a low-cost, locally manufactured smart electronic device. Specially developed for the measurement of atmospheric pollutants responsible for poor air quality, the developed device consists of low-cost components namely Arduino microcontroller board, toxic gases (ozone O_3, particle matter $PM_{2.5}$, carbon monoxide CO, carbon dioxide CO_2 and nitrogen dioxide NO_2), temperature (T) and, relative humidity (RH) sensors and XBee-based Internet of Things (IoT) wireless transmission modules. Two weeks of measurements were carried out in each hospital. The periodic average values obtained are 0.96 ± 0.06 ppm and 0.37 ± 0.09 ppm for O_3, $39.66 \pm 10\ \mu g/m^3$ and $39.72 \pm 10\ \mu g/m^3$ for $PM_{2.5}$, 0.41 ± 0.01 ppm and 0.42 ± 0.02 ppm for CO, 316.55 ± 63 ppm and 305.84 ± 89 ppm for CO_2, and 0.43 ± 0.01 ppm and 0.45 ± 0.02 ppm for NO_2 respectively in Biyem-assi hospital and Central hospital. The innovative prototype developed is simple and easy to implement with the advantage of measuring air pollution in real-time in order to assess health effects and environmental impact.

J. M. Taamté (✉) · N. Nducol · G. Ahmadou · R. C. E. Essama · M. K. Noubé · Y. F. T. Siaka · Saïdou
Research Centre for Nuclear Science and Technology, Institute of Geological and Mining Research, P.O. Box 4110, Yaoundé, Cameroon
e-mail: mtjfirst@yahoo.fr; taamtej@gmail.com

Saïdou
Nuclear Physics Laboratory, Faculty of Science, University of Yaoundé I, P.O. Box 812, Yaoundé, Cameroon

Keywords

Sensors · Smart electronic device · Air quality monitoring · Health risk assessment

1 Introduction

The many contaminants that impact indoor air quality (IAQ) include allergens, particles, carbon dioxide (CO_2), volatile organic compounds (VOCs), carbon monoxide (CO), formaldehyde (CH_2O), nitrous oxide (N_2O), glutaraldehyde ($C_5H_8O_2$), and bioaerosols. Patients with compromised immune systems, medical staff, and guests who spend a lot of time in a hospital setting may be especially exposed (Nimra et al., 2021).

Continuous automatic Monitoring of air quality is necessary to obtain up-to-dateinformation (Mahbub et al., 2020). When airborne pollution levels start to approach World Health Organisation (WHO) recommended limits, industries and environmental policy makers must act to cut emissions (Bell et al., 2011; Bruce et al., 2000). To better help them reduce emissions of air pollutants, two developed identical air quality measurement systems are manufactured locally and consist of a set of microcontroller board, XBee S2C wireless transmission modules, and air pollutant sensors whose reliability has been confirmed (Jacob et al., 2022). The sample was operated in two different hospitalization rooms at the Biyem-assi hospital and two more rooms at the main hospital in Yaoundé. Data on dangerous air pollutants, including ozone (O_3), particulate matter ($PM_{2.5}$), carbon monoxide (CO), carbon dioxide (CO_2), and nitrogen dioxide (NO_2), as well as temperature (T) and relative humidity (RH), were collected during the testing period. Real-time data collection was visualized using the XCTU PC application. Data collection and analysis were done to evaluate the exposure risk assessment. Additionally, the outcomes were compared to the WHO's international acceptable standards (Jacob et al., 2022).

K. W. Fomba et al. (eds.), *Advances in Air Quality Research in Africa*,
Advances in Science, Technology & Innovation, https://doi.org/10.1007/978-3-031-53525-3_7

2 Material and Method

2.1 Low-Cost Electronic Components

The electronic gas measuring sensors used react to toxic gases such as nitrogen dioxide (NO_2), carbon monoxide, carbon dioxide, ozone, suspended particulate matter ($PM_{2.5}$), as well as temperature and relative humidity. The MQ131 sensor has a long service life and is used for ozone measurement in a wide range. The CJMCU-6814 used for the nitrogen dioxide (NO_2) and carbon monoxide (CO) detection, is composed of three sensor chips with independent heating elements and sensitive layers. The DSM501B $PM_{2.5}$ sensor measures particles of size 2.5 μm ($PM_{2.5}$) in mg/m^3 which can be converted to $\mu g/m^3$, it is widely used for reliable data on air quality. The DHT22 sensor is composed of a thermistor temperature sensor and a capacitive humidity sensor. It is a calibrated digital output signal sensor. XBee S2C modules use the ZigBee radio frequency protocol based on the IEEE 802.15.4 standard with 2.4 GHz operating frequency. They provide IoT operation between electronic devices and computer through the XCTU application. The electronic card from the Arduino company is the Uno card whose heart is an Atmel ATMega328 microcontroller. This microcontroller is from the AVR family, programmable in C/C++ language. Table 1 gives the technical characteristics of the electronic sensors used.

2.2 Sensor Calibration

Sensors calibration and configuration of IoT modules were carried out before their usage in this work. Each sensor has been calibrated following the conditions and procedure given by the manufacturer with the exception of the DHT22 which is a digital sensor. It can be used without being recalibrated. The calibration of the MQ131 and CCS811 sensors was carried out according to the pre-calibration algorithm given in Jacob et al. (2021). The DSM501B dust sensor was also calibrated following the procedure described in Jacob et al. (2021); Yu and Zhang (2019). For the calibration of the CJMCU-6814 sensor (NO_2 and CO), the Arduino IDE code offered by the manufacturer was used to obtain the sensor calibration information. The recalibration is then carried out in an air-conditioned room over a period of half an hour.

2.3 Smart Electronic Devices Development

The proposed device is composed of an Arduino board, XBee S2C wireless transceiver modules, a power system, a 20 × 4 LCD display, a micro-SD card module, and several sensors including an ozone sensor (MQ131), a sulfur dioxide sensor (SPEC), a carbon monoxide and nitrogen dioxide sensor (CJMCU-6814), humidity and temperature sensor (DHT22), as well as the dust sensor (DSM501B). It is then composed of electronic sensors in contact with atmospheric pollutants, a central processing unit which is the Arduino microcontroller card, and output devices. Among the output devices is the data transmission module to a remote PC to provide the IoT function in this paper. A schematic diagram and final embodiment of the wireless communication device and its constituent elements are shown in Fig. 1.

These devices mainly measure pollutants responsible for air quality, as recommended by the World Health Organization and international air quality monitoring agencies like ATMO (France), USA, Canada, China, Belair (Belgium) etc. They also measure parameters including temperature (T) and relative humidity (RH) that are used to assess the presence of mold. The final embodiment of the device is shown in Fig. 2.

Table 1 Characteristics of used electronic sensors

Types of sensors	Measured gas	Detection capability	Sensor accuracy	Operating energy (V)
MQ131	O_3	0–1,000 ppb	±3%	5 ± 0.1
CCS811	CO_2	0–1000 ppm	±3%	5 ± 0.1
CJMCU-6814	NO_2	0–10 ppm	±5%	4.9–5.1
	CO	0–1,000 ppm	±10%	
SPEC sensor	SO_2	0–20 ppm	±2%	2.7–3.3
PPD42NS	$PM_{2.5}$	0–28,000 pcs/liter	±10%	5.0
DHT22	Temperature	−40–100 °C	±0.5 °C	3.3–6
	Humidity	0–100%	±2%	

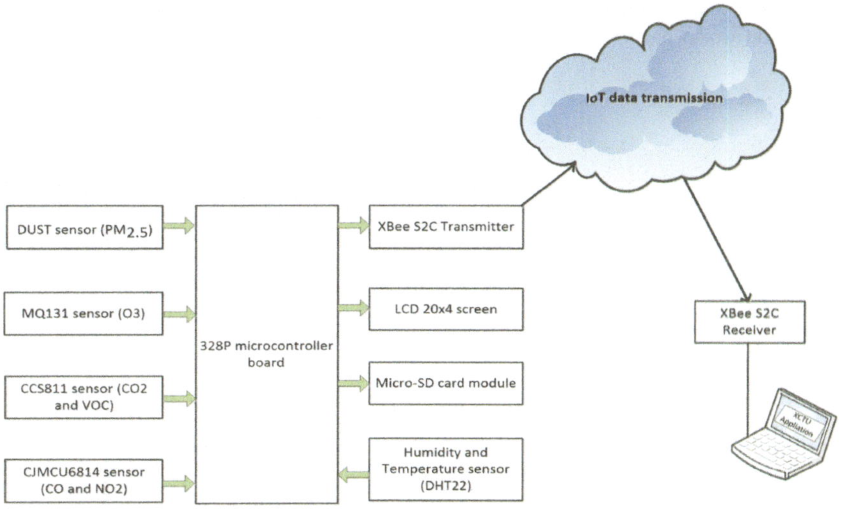

Fig. 1 Schematic diagram of the developed indoor air quality device

3 Results and Discussion

3.1 Atmospheric Data Sampling

There are many air pollutants which come from almost all human activities. They are transformed into synthetic molecules; either they remain in their raw state, burnt, or in the form of gas or matter particles. The air is unfortunately no longer pure. It contains a very complex mixture of all these pollutants, in varying concentrations depending on where you are and the season. Pollutants such as $PM_{2.5}$, NO_2, O_3, SO_2, and CO have adverse effects on human and environmental health (Jacob et al., 2022). It is recommended by the WHO to measure these gases to properly assess the air quality in a locality (Kim et al., 2003).

The concentrations of the pollutants O_3 (Fig. 3a), suspended particles (Fig. 3b), CO (Fig. 3c), CO_2 (Fig. 3d) and NO_2 (Fig. 3e) were presented in Fig. 3. Variations in indoor air pollution in the two hospitals were observed through the evolution of daily average values. In the light of Fig. 3, the histograms in red represent the average data of Biyem-assi hospital and the histograms in blue the average of the daily values obtained at the central hospital.

3.2 Health Risk Assessment

Two approaches of air quality assessment exist: monitoring air quality through direct (portable devices) or indirect (station-based monitors) measurements, and the approach through biological measurements using biological markers.

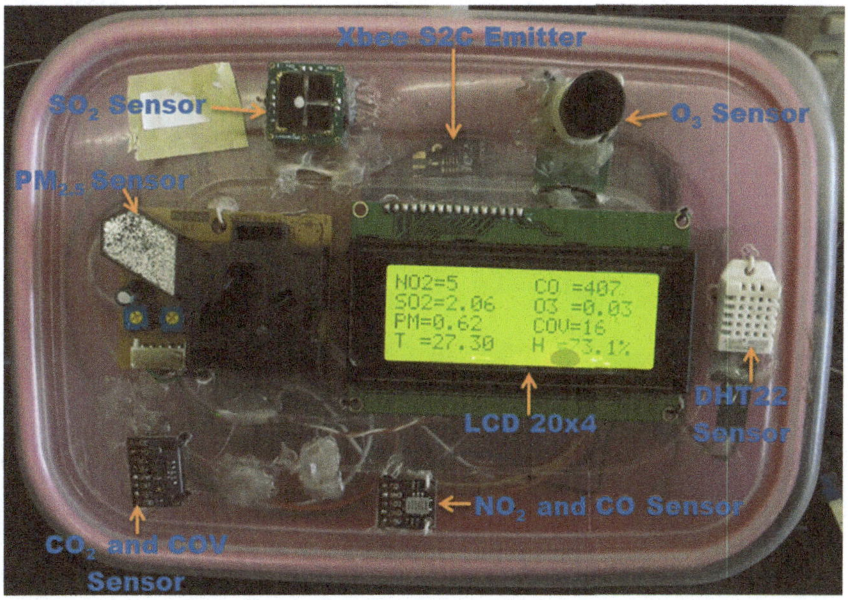

Fig. 2 IoT developed air quality monitoring device

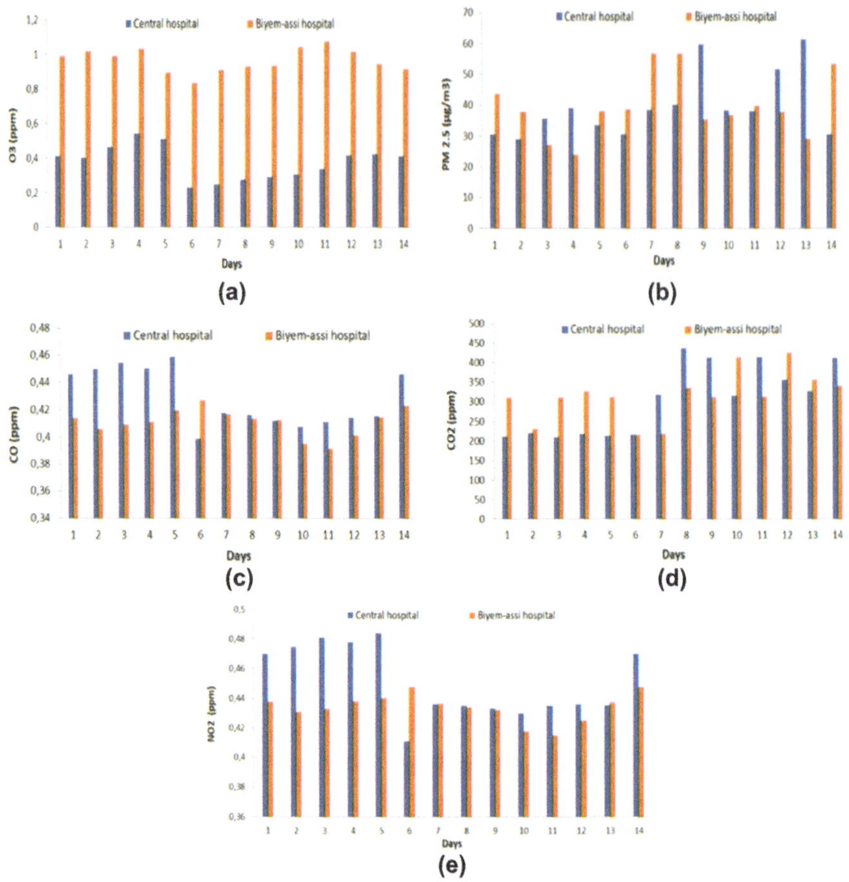

Fig. 3 Temporal evolution of the data in the two hospitals

In this article, air quality monitoring is based on direct measurements with locally manufactured monitors. Thus, analysis and interpretation of the data obtained and their comparison with the recommended threshold values are necessary. The recommended thresholds for the pollutants measured are those of the WHO given in Table 2. These values made it possible to calculate the air quality sub-indexes of the hospitals using Eq. (1) and each locality's final air quality indices (AQI) is the highest value of all sub-indexes for that locality.

$$\text{IQA sub-index} = (\text{measure}/\text{reference value}) \times 50 \quad (1)$$

The Air Quality Index (AQI) is an information and awareness tool designed to inform the population about the air quality in their locality and its impact on health. The assessment of air quality consists of calculating the AQI (Jacob et al., 2022) and the Air Quality Health Index (AQHI). The reference AQIs presented in Table 3 are used to assess AQIs in all localities around the world, in order to know whether

Table 2 Pollutants responsible for air quality and their WHO reference values

Pollutants	WHO guideline values	Health effects
Particles with a diameter less than 2.5 µm ($PM_{2.5}$)	15 µg/m^3 daily average 5 µg/m^3 annual average	Risks of developing cardiovascular and respiratory diseases
Ozone (O_3)	100 µg/m^3 on average over 8 h	May induce breathing difficulties and asthma. Risk of disruption of the functioning of the lungs
Nitrogen dioxide (NO_2)	10 µg/m^3 annual average 25 µg/m^3 hourly average	Risk of development of chronic bronchitis in asthmatic subjects
Sulfur dioxide (SO_2)	40 µg/m^3 daily average	May cause respiratory and pulmonary system function and eye irritation
Carbon monoxide (CO)	60 µg/m^3 daily average	Asphyxiant gas that attaches to red blood cells and prevents them from carrying oxygen properly in the body

Table 3 Standard air quality index and level of health impact (Bruce et al., 2000)

AQI values	Levels of health concern
From 0 to 50	Good
From 51 to 100	Moderate
From 101 to 150	Unhealthy for sensitive groups
From 151 to 200	Unhealthy
From 201 to 300	Very unhealthy
From 301 to 500	Hazardous

Table 4 Air quality health index (Jacob et al., 2021; Wong et al., 2013)

Helth risk category	Low	Moderate	High	Very high	Serious
$AQHI_{2.5}$ / $AQHI_{10}$	From 0 to 3	From 4 to 6	To 7	From 8 to 10	Beyond 10

the air quality is good or poor (Jacob et al., 2022; Olstrup, 2020; Szyszkowicz, 2019; Wong et al., 2013).

The data obtained in Fig. 3 was converted to the appropriate units as in Table 2 and used to calculate the indoor AQI of each hospital. The AQI of the Central Hospital is 407.42 that of the Biyem-assi Hospital is 388.46 and the two values are inclusive in the interval from 301 to 500 in Table 3. This makes it possible to affirm that in general, the quality of air is hazardous for the sick and sensitive people in these hospitals due to the high concentrations of CO in the environment (Jacob et al., 2022).

The AQHI, whose assessment intervals are given in Table 4, is the risk of population exposure to air pollutants (Gryech et al., 2020; Kuzum et al., 2013). It is evaluated based on three main pollutants, namely ozone, fine particles $PM_{2.5}$ ($AQHI_{2.5}$), and nitrogen dioxide (NO_2). The evaluation of $AQHI_{2.5}$, described by Jacob et al., in Jacob et al. (2022) is obtained based on Eq. (2):

$$AQHI_{2.5} = \frac{10}{10.4} \times \left[100 \times \left(\left(e^{0.000871 * NO_2} - 1\right) + \left(e^{0.000537 \times O_3} - 1\right) + \left(e^{0.000487 \times PM_{2.5}} - 1\right)\right)\right] \quad (2)$$

NO_2 and O_3 in ppb and $PM_{2.5}$ in µg/m^3. The $AQHI_{2.5}$ obtained is compared to the results in Table 3 to determine the level of exposure risk associated with the air of each hospital.

By using data obtained in Fig. 3, the $AQHI_{2.5}$ of the Central hospital is 69.15 and that of the Biyem-assi hospital is 111.28 and these two values are beyond 10, therefore, the unhealthy air in the two hospitals has a strong impact on the health of patients, sensitive people and the majority of staff who are exposed every day (Jacob et al., 2022).

4 Conclusion

This work made it possible to carry out real-time monitoring of air quality in health establishments using low-cost electronic devices produced locally. The measurement campaigns carried out in hospitals from October 15 to 30, 2021 at Biyem-assi Hospital and from September 1 to 15, 2021 at the Central Hospital made it possible to obtain 0.96 ± 0.06 ppm and 0.37 ± 0.09 ppm of O_3, 39.66 ± 10 µg/m^3 and 39.72 ± 10 µg/m^3 of $PM_{2.5}$, 0.41 ± 0.01 ppm and 0.42 ± 0.02 ppm of CO, 316.55 ± 63 ppm and 305.84 ± 89 ppm of CO_2, and 0.43 ± 0.01 ppm and 0.45 ± 0.02 ppm of NO_2 respectively in Biyemassi hospital and central hospital and the assessment of risks exposure through the AQI and the $AQHI_{2.5}$. It has been found that people are exposed to unhealthy air and the risk of exposure is dangerous for sensitive people and low for other people. This equipment leverages low-cost sensors for air sampling and health risk assessment. It is a decision-making tool for air quality and health risk assessment. We will work in the future on improving the functionality of the device in order to meet the challenge of air pollution.

References

Bell, M. L., Cifuentes, L. A., Davis, D. L., Cushing, E., Telles, A. G., & Gouveia, N. (2011). Environmental health indicators and a case study of air pollution in Latin American cities. *Environmental Research, 111*(1), 57–66.

Bruce, N., Perez-Padilla, R., & Albalak, R. (2000). Indoor air pollution in developing countries: A major environmental and public health challenge. *Bulletin of the World Health Organization, 78*(9), 1078–1092.

Gryech, I., Ben-Aboud, Y., Guermah, B., Sbihi, N., Ghogho, M., & Kobbane, A. (2020). MoreAir: A low-cost urban air pollution monitoring system. *Sensors, 20*(4), 998.

Jacob, M. T., Nasser, N., Michaux, K. N., Flore, Y., & Siaka, T. (2022). Zigbee-based wireless smart device for enclosed space real-time air quality monitoring: Experiment, data analysis and risk assessment. *Indoor Air Quality Assessment for Smart Environments, 30*, 55.

Jacob, M. T., Michaux, K. N., Bertrand, B., Yvette Flore, T. S., Nasser, N., Vitrice Ruben, F. S., ... & Saïdou. (2021). Low-cost air quality monitoring system design and comparative analysis with a conventional method. *International Journal of Energy and Environmental Engineering, 12*(4), 873–884.

Kim, C. W., Phipps, T. T., & Anselin, L. (2003). Measuring the benefits of air quality improvement: A spatial hedonic approach. *Journal of Environmental Economics and Management, 45*(1), 24–39.

Kuzum, D., Yu, S., & Wong, H. P. (2013). Synaptic electronics: Materials, devices and applications. *Nanotechnology, 24*(38), 382001.

Mahbub, M., Hossain, M. M., & Gazi, M. S. A. (2020). IoT-Cognizant cloud-assisted energy efficient embedded system for indoor intelligent lighting, air quality monitoring, and ventilation. *Internet of Things, 11*, 100266.

Nimra, A., Ali, Z., Nasir, Z. A., Tyrrel, S., & Sidra, S. (2021). Characterization of indoor air quality in relation to ventilation practices in hospitals of Lahore Pakistan. *Sains Malaysiana, 50*(6), 1609–1620.
Olstrup, H. (2020). An air quality health index (AQHI) with different health outcomes based on the air pollution concentrations in Stockholm during the period of 2015–2017. *Atmosphere, 11*(2), 192.
Szyszkowicz, M. (2019). The Air Quality Health Index and all emergency department visits. *Environmental Science and Pollution Research, 26*, 24357–24361.
Wong, T. W., San Tam, W. W., Yu, I. T. S., Lau, A. K. H., Pang, S. W., & Wong, A. H. (2013). Developing a risk-based air quality health index. *Atmospheric Environment, 76*, 52–58.
Yu, T., & Zhang, T. Q. (2019). Design and implementation of a $PM_{2.5}$ dust detection system based on ZigBee. In *2019 International Conference on Electronic Engineering and Informatics (EEI)* (pp. 260–264). IEEE.

Ambient Air Quality Characterisation in a Farming Town on the South African Highveld

Tonderai Dangare and Newton R. Matandirotya

Abstract

Ambient air pollution remains a leading contributor towards the burden of diseases across the world. The purpose of this study was to characterise the quality of ambient air over a farming community of Standerton, located on the South African Highveld, Mpumalanga over the year 2020. The study focused on four pollutants namely NO_2, O_3, $PM_{2.5}$ and PM_{10}. Findings were that PM_{10} and $PM_{2.5}$ ambient concentrations were elevated during the winter season while NO_2 and O_3 ambient concentrations got elevated during spring and summer respectively. The study attributes the high particulate matter during winter to domestic solid fuel burning and biomass burning. On the other hand, the elevated ambient O_3 during summer was linked to increased photochemical activities during the summer season. As a farming community, there are a number of policy options available to reduce ambient air pollution and these include curbing biomass burning and a reduction in reliance on solid fuels.

Keywords

Air pollution · Burden of disease · Particulate matter · Air quality · Solid fuel

T. Dangare
Department of Electronics and Telecommunications, University of Zimbabwe, 630 Churchill Ave Mt Pleasant, Harare, Zimbabwe
e-mail: tdangare@ceic.uz.ac.zw

N. R. Matandirotya (✉)
Department of Geosciences, Faculty of Science, Nelson Mandela University, Port Elizabeth 6000, South Africa
e-mail: runyamore@gmail.com

Climate Change and Resilience Centre, Kgotso Development Trust, P.O Box 5, Beitbridge, Zimbabwe

1 Introduction

Poor air quality has shown a statistically significant correlation with increasing cardiovascular and respiratory-related ailments such as heart diseases, stroke, asthma and premature deaths around the world (Huang et al., 2021). Particulates of aerodynamic diameter measuring below 2.5 μm ($PM_{2.5}$) possess the most health-damaging effects on human health as they can penetrate deep into the human respiratory system potentially leading to respiratory infections (Etim et al., 2021). On the other hand, NO_2 also has some negative human health impacts that include coughing, wheezing and also worsening those with asthmatic conditions. NO_2 plays a big role in the chemical formation of both surface and tropospheric ozone (O_3) which is a greenhouse gas (Wang et al., 2020).

The Highveld of South Africa is identified as having the poorest air quality in the country. This highly industrialized area that includes Gauteng and Mpumalanga provinces also consists of low-income residential areas. In Mpumalanga, several studies carried out to analyze air pollution trends have identified domestic fuel burning and coal-fired power plants as major contributors to air pollution. Although several studies have been done on air quality over Mpumalanga Highveld, Standerton has not been covered.

The current study characterizes ambient air over the farming community of Standerton, Mpumalanga Highveld in South Africa during the year 2020 with the aim of filling pollution data gaps. Standerton is a farming community however it also has some background industrial sources as well as coal-fired thermal power plants. Our study investigated four ambient pollutants namely NO_2, SO_2, $PM_{2.5}$ and PM_{10} over the year 2020.

K. W. Fomba et al. (eds.), *Advances in Air Quality Research in Africa*,
Advances in Science, Technology & Innovation, https://doi.org/10.1007/978-3-031-53525-3_8

2 Materials and Methods

Pollution data was retrieved from the South African Air Quality Information System (SAAQIS) (SAAQIS, 2023). SAAQIS is an online platform that hosts air pollution and meteorological data from monitoring stations in South Africa. Retrieved data were cleaned and quality checked for instrument errors. Analysis of data was done using R openair statistical package. Plots were used to establish air pollution levels, variation patterns and also to investigate the influence of wind direction and speed on pollutant concentrations during the different seasons of the year.

3 Results and Discussion

The time series plot for the pollutants measured in Standerton Highveld South Africa as well as the air quality guideline limits for $PM_{2.5}$ and PM_{10} according to WHO standards (WHO, 2021) are shown in Fig. 1. The plot shows a general trend of pollutants peaking concentration levels during the winter season (JJA).

The monthly time variation plots of both the particulate matter (Fig. 2a(ii)) and gaseous pollutant (Fig. 2b(ii)) reveal pollutant level characteristics observed in Fig. 1 where emission levels peak during winter. The weekday cycles (Fig. 2a(iii), b(iii)) show that emission levels tend to drop during weekends, this is an emission characteristic of industrialized environments where emissions abate due to factory closure during weekends. The diurnal plots (Fig. 2a(i), b(i)) show two emission peaks for both particulate matter and gaseous pollutants that can be attributed to morning and evening rush hours.

Bivariate polar plots (Fig. 3) show the dependence of particulate matter levels on wind direction and speed at the monitoring site. Wind speed is zero at the center of each plot, increasing radially outward while levels of concentrations are scaled according to the colour on the key. Plots were 'conditioned' according to the season of the year. For both $PM_{2.5}$ and PM_{10} (Fig. 3(i–ii)) there was concentration elevation at wind speeds above 8 ms^{-1} blowing from westerly and northwesterly directions during winter. High wind speeds can give rise to the resuspension of particulate matter from sand and dust storms giving rise to elevated PM_{10} pollution levels.

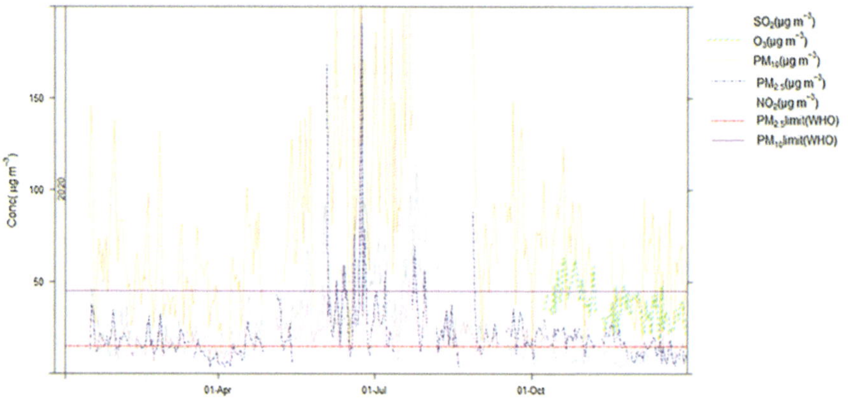

Fig. 1 Time series plot of pollutants measured in Standerton and the WHO daily averaged air quality guideline limits for: $PM_{2.5}$ (horizontal dashed red line) and PM_{10} (horizontal pink line)

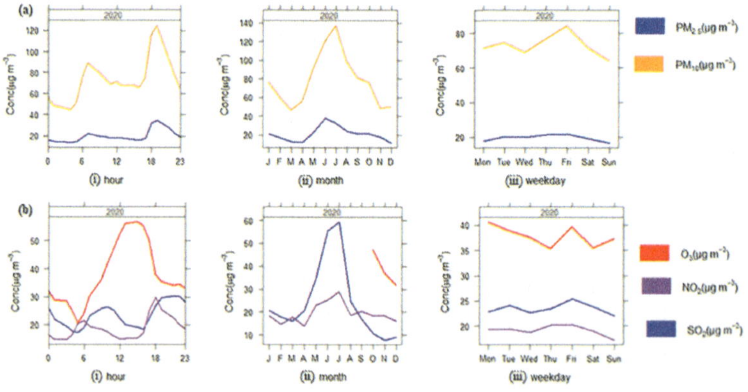

Fig. 2 Concentration variation plots for hourly (i), monthly (ii) and according to the day of the week (iii) averages. The upper raw (**a**) is for particulate matter whereas the lower raw (**b**) is for gaseous pollutants

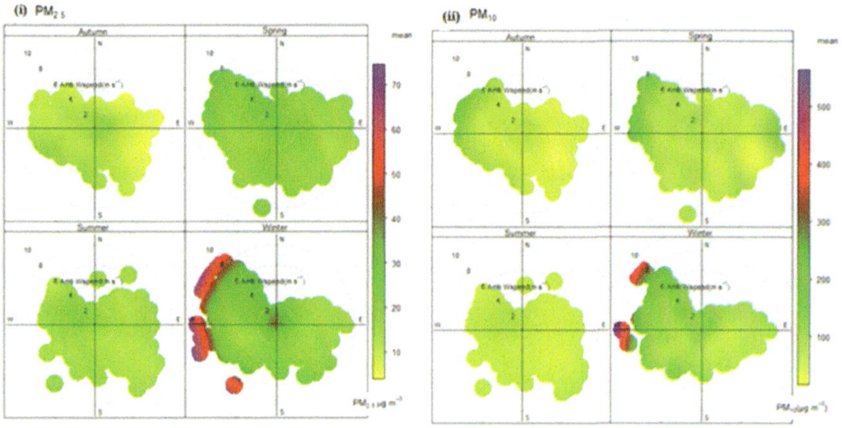

Fig. 3 Polar plots for PM concentrations (μg m^{-3}) during the year 2020 in Standerton

4 Conclusion

The current study characterized the quality of ambient air during the year 2020 in Standerton. Elevated concentration levels of all measured pollutants were observed during the winter season. High $PM_{2.5}$ and PM_{10} concentrations levels were observed to be associated with wind speeds above 8 ms^{-1} blowing from an easterly direction during winter. Analysis of time series plots revealed an acute emission episode of all pollutants in the month of July lasting for about three days. During this period all pollutants were highly elevated. Air pollution levels in the study area were characterized as largely higher than the WHO air quality guidelines. Anthropogenic activities and background industrial emissions in the agricultural area of Standerton were the suspected sources of the observed high levels of pollution.

References

Etim, M. A., Babaremu, K., Lazarus, J., & Omole, D. (2021). Health risk and environmental assessment of cement production in Nigeria. *Atmosphere*, *12*(9). MDPI, Sep 01. https://doi.org/10.3390/atmos12091111

Huang, S., et al. (2021). Long-term exposure to nitrogen dioxide and mortality: A systematic review and meta-analysis. *Science of the Total Environment*, *776*(February), 145968. https://doi.org/10.1016/j.scitotenv.2021.145968

SAAQIS. Retrieved September 07, 2023, from https://saaqis.environment.gov.za/

Wang, Z., Anthony, J. L., Erickson, L. E., Higgins, M. J., & Newmark, G. L. (2020). Nitrogen dioxide and ozone pollution in the Chicago metropolitan area. *Journal of Environmental Protection (Irvine, Calif)*, *11*(08), 551–569. https://doi.org/10.4236/jep.2020.118033

WHO. (2021). Global air quality guidelines. Particulate matter (PM2.5 and PM10), ozone, nitrogen dioxide, sulfur dioxide and carbon monoxide. Geneva: World Health Organization.

Impact of Mineral Dust on Air Quality: Case Study from Cabo Verde

Daniel Tetteh Quaye, Khanneh Wadinga Fomba, Nongma Kaboré and Hartmut Herrmann

Abstract

Mineral dust negatively impacts air quality and has been linked to health risks and mortality. This study quantifies the contribution of mineral dust to air pollution in São Vicente, Cabo Verde. A total of 129 particulate matter (PM_{10}) filter samples were collected at a strategic site using a high-volume (HV) sampler in 2018. The mass concentration of the particles was determined using gravimetry. The results showed that the average annual PM_{10} concentration (43 µg/m^3) in São Vicente, Cabo Verde, was more than twice the air quality guideline defined by the World Health Organization (WHO) and the European Union (EU). Additionally, 25% of the PM_{10} samples exceeded the WHO daily PM_{10} threshold (50 µg/m^3), with dust outbreaks responsible for 34% of these samples. Overall, dust influence represented 47% of PM_{10} mass concentration throughout the year. A strong temporal variation in PM_{10} mass concentration was observed. The highest concentration was in January (90.7 ± 77.4 µg/m^3) and the lowest in May (18.7 ± 4.3 µg/m^3). Elevated PM_{10} concentrations indicate that public health may be at risk from high PM_{10} exposure.

Keywords

Mineral dust · PM_{10} · Air quality · Cabo Verde

D. T. Quaye · N. Kaboré
West African Science Service Centre on Climate Change and Adapted Land Use (WASCAL), Universidade Técnica do Atlântico (UTA), Campus de Ribeira de Julião, CP 163 Mindelo, São Vicente, Cabo Verde
e-mail: dquaye@uta.cv

K. W. Fomba · H. Herrmann (✉)
Atmospheric Chemistry Department (ACD), Leibniz Institute for Tropospheric Research (TROPOS), Permoserstraße 15, 04318 Leipzig, Germany
e-mail: herrmann@tropos.de

1 Introduction

In recent years, particulate matter (PM) concentrations in the atmosphere have received increased attention due to their associated human health risks. Both natural and anthropogenic sources influence the amount of PM in the atmosphere (Gama et al., 2020).

Amongst the natural sources, mineral dust—mainly from the Sahara Desert—is important and has been linked to varied health effects and mortality (Bouet et al., 2019). Mineral dust impacts air quality by increasing the levels of atmospheric PM. Despite its critical nature, mineral dust impact studies in Africa are limited.

In São Vicente (Cabo Verde), air quality is influenced by both natural (Saharan Dust) and local anthropogenic emissions (Almeida-Silva et al., 2013). Unlike anthropogenic emissions, natural emissions can only be assessed but not controlled (Gama et al., 2020). As confirmed by Almeida-Silva et al. (2013), PM levels are regularly driven beyond established standards in such a remote region frequently impacted by dust. Therefore, it is essential to constrain mineral dust contribution from local emissions.

This study aims to quantify the impact of mineral dust on particle mass concentration in São Vicente and, in turn, air quality. One year (2018) of PM samples collected at the Cape Verde Atmospheric Observatory (CVAO) were used in quantifying the impacts. The CVAO, located in a remote area in São Vicente, is barely affected by human emissions and offers a unique setting to monitor dust's effect on air quality. Information from the study will be crucial to future PM guidelines and efforts to reduce anthropogenic emissions, especially if benefits are to be derived.

K. W. Fomba et al. (eds.), *Advances in Air Quality Research in Africa*,
Advances in Science, Technology & Innovation, https://doi.org/10.1007/978-3-031-53525-3_9

2 Materials and Methods

2.1 Study Area and Sampling Site

São Vicente is one of the islands of the Cabo Verde archipelago. The archipelago lies in the path of Saharan dust transport to the Atlantic Ocean (Freire et al., 2020). The region is arid and semi-arid, dry and hot, with scarce rainfall and a yearly mean temperature of 25 °C. In Cabo Verde, São Vicente has the highest population density at 361 inhabitants/km^2 (Freire et al., 2020).

The Cape Verde Atmospheric Observatory (16°51.490 N, −24°52.020 E) is situated in the eastern part of São Vicente. The site is 900 km from the West African coast, 70 m from the shoreline and 10 m a.s.l. Due to its location, the site experiences strong northeasterly trade winds and negligible influence from local emissions (Fomba et al., 2014). Particles arriving here are therefore driven by long-range transport.

2.2 Sample Collection and Analysis

Particulate matter sampling was carried out in 2018 at the top of a 30 m measurement tower at CVAO. This was to limit the effect of sea spray on sampled particles. The samples were collected on quartz fibre filters (Munktell, MK 360) using a high volume (HV) PM_{10} Digitel (DHA-80) collector. A 72-h sampling timeframe was adopted except during rigorous campaigns when this was adjusted to 24 h.

The sampled filters were weighed with a microbalance (Mod. AT261 Delta Range, Mettler-Toledo) after being stabilized for three days at a constant temperature of 20 ± 1 °C and relative humidity of 50 ± 5% before and after sampling. The weight difference was calculated and divided by the total sampling volume to derive the mass concentration.

3 Results and Discussion

3.1 PM_{10} Mass Concentration

To quantify the contribution of dust to air pollution in São Vicente, PM_{10} was collected in 2018. A total of 129 samples were obtained. At the CVAO, PM_{10} concentration is a good index for mineral dust (Fomba et al., 2014). The annual mean PM_{10} concentration determined for the period was 43 µg/m^3. Within the region, this value was similar to the means of 47.1 and 49.7 µg/m^3 reported by Fomba et al. (2014) and Gama et al. (2020), respectively. But, unlike this study and the referenced studies, Almeida-Silva et al. (2013) recorded a mean of 60 µg/m^3 in Santiago, another island of Cabo Verde. The difference could be due to local emissions impacting PM collection at their sampling location.

In comparison to air quality guidelines, the annual mean was more than double the 20 µg/m^3 annual limit set by the World Health Organization (WHO, 2006) and the European Union (De Longueville et al., 2010). But it was just below the 50 µg/m^3 benchmark established by the United States Environmental Protection Agency (US-EPA) (De Longueville et al., 2010).

3.2 Variation of PM_{10} Concentrations

The time series of PM_{10} mass concentration over the study period shows a strong seasonal variation (Fig. 1).

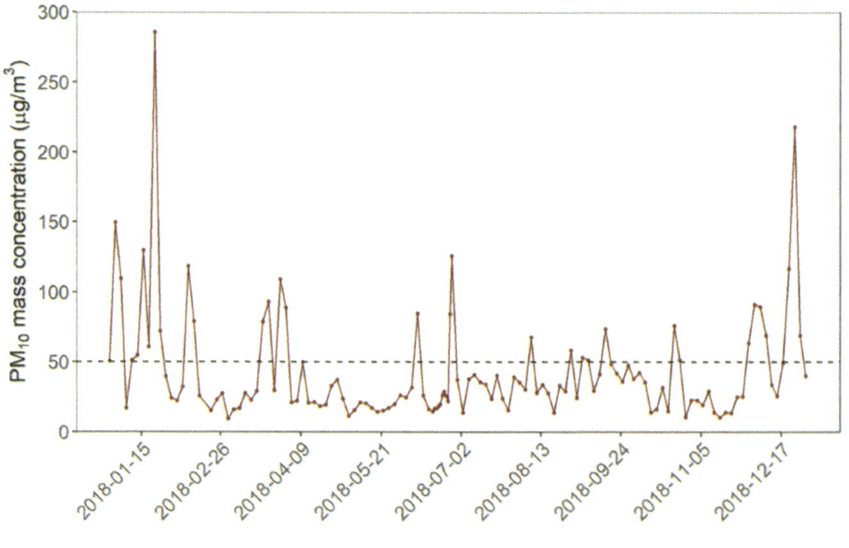

Fig. 1 Time series of PM_{10} mass. Dashed line represents WHO's 24-h mean PM_{10} limit

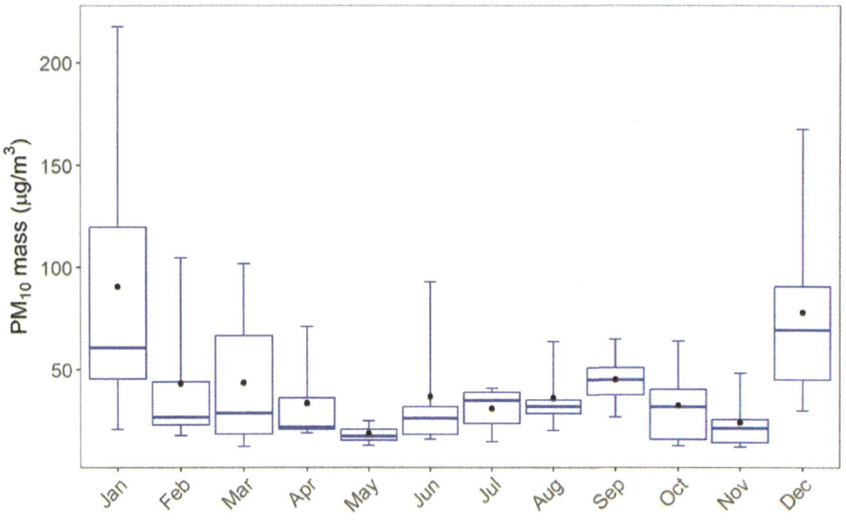

Fig. 2 Seasonal variation of PM_{10} mass in 2018. The black dot inside the box represents the mean, whiles the horizontal line shows the median value. The top and bottom of each box are the 25th and 75th percentiles. The whiskers are the 5th and 95th percentiles

Maximum and minimum monthly average concentrations were recorded in January (90.7 ± 77.4 µg/m^3) and May (18.7 ± 4.3 µg/m^3), respectively (Fig. 2). The highest concentrations were recorded during the winter months, with a sharp drop noticeable in February (Fig. 1). During winter, Saharan dust is transported over the Atlantic at lower altitudes with the trade winds (Fomba et al., 2013).

The observed average mass concentrations for spring, autumn, summer, and winter were 31.5 ± 25.5, 33.8 ± 18.2, 35.0 ± 23.0 and 73.3 ± 61.4 µg/m^3, respectively. This trend is similar to that reported by Fomba et al. (2014) at the site, with a minor difference: autumn recorded a higher mean than summer, unlike in this study. This could be due to relatively higher episodic dust events observed in the summer.

According to Fomba et al. (2013), PM_{10} mass exceeding 30 µg/m^3 at CVAO indicates some degree of dust influence. As shown in Fig. 1, this situation describes 47% of the year 2018 (61 samples). Observations below 30 µg/m^3 are representative of remote background conditions (Fig. 1). In the study area, major Saharan dust outbreaks are characterized by observations exceeding 90 µg/m^3 (Fomba et al., 2014). Our results show that 25% of the PM_{10} samples exceeded the WHO 24-h mean PM_{10} limit (50 µg/m^3), with dust outbreaks responsible for 34% of these samples. The highest number of exceedances were observed during the winter (Table 1). The spikes in Fig. 1 depict these events.

Table 1 Seasonal contribution of dust outbreak to exceedances of WHO's daily mean PM_{10} limit. Number of samples with exceedances is N

	Winter	Spring	Summer	Autumn
N	17	4	5	6
Dust outbreak seasonal contribution (%)	47	50	20	0

4 Conclusion

This study quantified the impact of mineral dust on air quality in São Vicente, Cabo Verde using PM_{10} samples collected at the CVAO. Different degrees of dust concentration influence the area, describing 47% of PM_{10} mass concentration in 2018. The annual mean PM_{10} concentration of 43 µg/m^3 exceeded internationally established standards, though exclusive of local emissions. This could have health implications for the public. Consequently, we recommend epidemiological studies into the health effects of mineral dust exposure. Also, as this source cannot be controlled, anthropogenic emissions in the region need to be assessed if such information is not available. The information can be used to drive strong regulation through the definition of locally relevant PM_{10} thresholds for air quality.

Acknowledgements The authors would like to thank Luis Neves for sample collection at CVAO and the TROPOS laboratory assistants for their assistance. Financial support from the BMBF in the framework of the WASCAL Graduate Study Programme provided to Daniel Tetteh Quaye is also appreciated.

References

Almeida-Silva, M., Almeida, S. M., Freitas, M. C., Pio, C. A., Nunes, T., & Cardoso, J. (2013). Impact of Sahara dust transport on Cape Verde atmospheric element particles. *Journal of Toxicology and Environmental Health, 76*(4–5), 240–251.

Bouet, C., Labiadh, M. T., Rajot, J. L., Bergametti, G., Marticorena, B., Henry des Tureaux, T., & Féron, A. (2019). Impact of desert dust on air quality: What is the meaningfulness of daily PM standards in regions close to the sources? The example of Southern Tunisia. *Atmosphere, 10*(8), 1–22.

De Longueville, F., Hountondji, Y. C., Henry, S., & Ozer, P. (2010). What do we know about effects of desert dust on air quality and human health in West Africa compared to other regions? *Science of the Total Environment, 409*(1), 1–8.

Fomba, K. W., Müller, K., Van Pinxteren, D., & Herrmann, H. (2013). Aerosol size-resolved trace metal composition in remote northern tropical Atlantic marine environment: Case study Cape Verde islands. *Atmospheric Chemistry and Physics, 13*(9), 4801–4814.

Fomba, K. W., Müller, K., Van Pinxteren, D., Poulain, L., Van Pinxteren, M., & Herrmann, H. (2014). Long-term chemical characterization of tropical and marine aerosols at the Cape Verde Atmospheric Observatory (CVAO) from 2007 to 2011. *Atmospheric Chemistry and Physics, 14*(17), 8883–8904.

Freire, S. M., Relvas, H., Lopes, M. (2020) Impact of traffic emissions on air quality in Cabo Verde. *Environmental Monitoring and Assessment, 192*(11), 1–13.

Gama, C., Pio, C., Monteiro, A., Russo, M., Fernandes, A. P., Borrego, C., & Tchepel, O. (2020). Comparison of methodologies for assessing desert dust contribution to regional PM10 and PM2.5 levels: A one-year study over Portugal. *Atmosphere, 11*(2), 1–21.

World Health Organization. WHO air quality guidelines: Global update 2005—Particulate matter, ozone, nitrogen dioxide and sulfur dioxide; World Health Organization, Regional Office for Europe.

Novel Techniques in Quantifying Microplastics in Atmospheric Particles

Majda Mekic and Hartmut Herrmann

Abstract

The accumulation of plastic debris is characterized as one of the most growing and emerging environmental issues of the twenty-first century because there is no place on the Earth that is free from plastic pollution, even in remote places with very limited or none human activity. Even though the sources and sinks of plastic pollution are not well constrained, the fragmentation through mechanical action of wind and/or waves, and interaction with high solar irradiance combined with high temperatures could accelerate and promote the formation of microplastic particles (MP), with an aerodynamic diameter between 1 μm and 5 mm. The abundance of MP in the atmosphere is of great concern since they can be easily distributed within all environmental compartments and ingested by many organisms in the global ecosystem. Moreover, the possible harmful effects related to their biopersistant nature clearly indicate a need for further studies to access their atmospheric abundances. This requires the implementation of reliable sampling and analytical tools that can be provide information about the physical properties such as their size fractions and chemical composition at the molecular level. New techniques such as Curie-Point Pyrolysis GC MS are gaining admiration for such analysis as they can improve the atmospheric quantification of the MP at a molecular level, which is required to better constrain the MP sources and global plastic pollution cycle, currently plagued with large uncertainties due to data scarcity.

Keywords

Plastic aerosol · Atmospheric microplastics · Nanoplastic · Plastic pollution · Accumulation of plastic debris

1 Introduction

Plastic pollution is becoming one of the most growing environmental issues, since the majority of plastic precursors (monomers) originate from fossil hydrocarbons, which are non-biodegradable. Hence, their lifetime is expected to be significantly long, resulting in the accumulation of plastic debris rather than their decomposition (Barnes et al., 2009; Geyer et al., 2017). Additionally, a mixture of different chemicals such as dyes and plasticizers is involved in the synthesis of synthetic polymers, which could exhibit potential toxic properties (Rochman et al., 2019; Zimmermann et al., 2020). Following their persistent nature, plastic debris continuously fragment into smaller pieces due to the influence of mechanical forces, sunlight, temperature, pH, and wind or wave action, which would gain an increasing number of plastic particles with micro- (from 5 mm to 1 μm) and nano- (<1 μm) scaled aerodynamic diameter (Cai et al., 2018; Kale et al., 2015). Even though the synthetic polymers have already been identified in different matrices including surface waters (Jambeck et al., 2015; Sebille et al., 2016), marine biota (Ugwu et al., 2021), dust (Liu et al., 2019), and even human blood (Leslie et al., 2022) and feces (Schwabl et al., 2019), a tremendous knowledge gaps exist on ambient mass- and number-concentrations in atmospheric particles smaller than 10 μm in diameter. The surge in ambient MP investigation indicates the curiosity in understanding the fate of MP ambient exposure and in closing the existing knowledge gabs about the levels and environmental impacts. Recent modelling studies (Brahney et al., 2021; Lebreton & Andrady, 2019) have called for in-situ determination of MP in the atmosphere since the

M. Mekic · H. Herrmann (✉)
Atmospheric Chemistry Department (ACD), Leibniz-Institut for Tropospheric Research (TROPOS), Permoserstraße 15, 04318 Leipzig, Germany
e-mail: herrmann@tropos.de

K. W. Fomba et al. (eds.), *Advances in Air Quality Research in Africa*,
Advances in Science, Technology & Innovation, https://doi.org/10.1007/978-3-031-53525-3_10

experimental database is scarce or do not exist at all. This knowledge is needed for a better understanding of the airborne MP sources and emission mechanisms on a global scale. Furthermore, atmospheric aging processes of synthetic polymers, formation of their degradation products, followed by atmospheric multiphase processing with common gaseous or particulate compounds and their subsequent influence on the air quality is largely lacking. Modelling results (Brahney et al., 2021; Lebreton & Andrady, 2019) suggest that the marine environment could account as a potential source of airborne MP since those insoluble particles could be easily released into the atmosphere via wave or wind action. Tire wear particles, released upon the mechanical process of breaking a vehicle and road emissions could account for a significant source of elastomer-based airborne particles. Additionally, agricultural fields are recognized as an important source of dust-containing MP particles, which could be associated with unregulated biomass burning facilities, especially in developing countries (Brahney et al., 2021). Hence, the fate of MP in the atmosphere needs further investigations to tackle their emission sources more accurately and to identify their atmospheric lifetime while keeping in mind their unusual shapes (long fibers). Since airborne synthetic polymer particles, having an aerodynamic diameter smaller than 20 μm are considered respirable, detection of their trace amount is of significant importance to confirm exposure levels, needed to develop public health protection measures (Leslie & Depledge, 2020). MP particles, larger than 10 μm or 50 μm have already been identified and quantified with existing analytical methods, among which the promising results were gained by spectroscopic techniques, like μFT-IR is (Veerasingam et al., 2021). In addition, Kernchen et al. (2022) investigated the MP in rural, suburban and urban aerosol samples by implementing Raman spectroscopy.

The determined size distribution of the airborne MP is depicted on Fig. 1, where the length of the examined airborne MP ranged between 11 μm and 100 μm for 90% of the investigated samples. Furthermore, the results from Fig. 1 reveals that the relative abundances of particles decrease with the increase of particle diameter. Apart from the fact that the finding overlaps with already existing literature, the determined number concentration of MP might be overestimated due to the size limitation of the applied analytical technique, which is providing an infrared spectrum of absorption or emission of an investigated synthetic polymer. Additionally, large particles in the upper μm or even mm range could be observed by particle counting practice using optical microscopy (Barbosa et al., 2020; Zhang et al., 2020). Among others Curie-Point Pyrolysis GC MS is a preferable analytical technique, enabling identification of an investigated polymer feature on a molecular level due to thermal degradation (Ivleva, 2021). Additionally, this method could serve not only for the identification of monomers of synthetic polymers but also it could distinguish possible degradation byproducts (Ivleva, 2021). The subsequent duration of Curie-Point Pyrolysis GC MS analysis is longer compared to spectroscopic techniques but Curie-Point Pyrolysis GC MS has an advantage of the simultaneous identification and quantification of different polymers where the pyrogram is used as a fingerprint of the respective polymer (Primpke et al., 2020). In the scope of our AirPlast project, we are going to focus on improving the Curie-Point Pyrolysis GC MS analytical method and its application for reliable detection and quantification of airborne synthetic particles, smaller than 10 μm, in different environmental compartments such as urban, rural, street dust, soil, marine surface microlayer and similar. Furthermore, the obtained results are going to be incorporated into chemistry transport model to better predict and

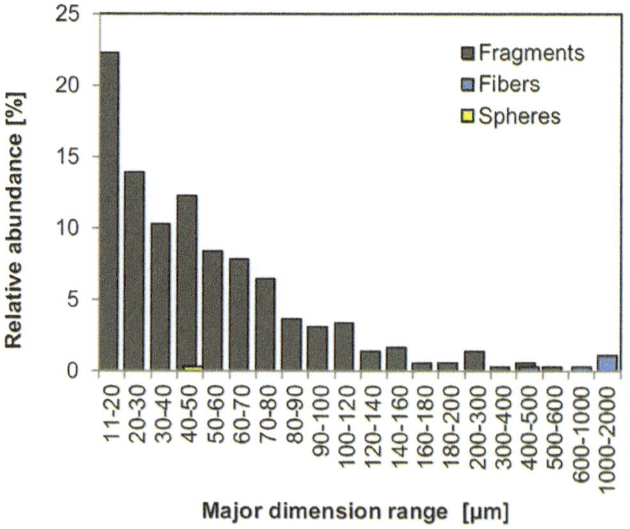

Fig. 1 Size distribution of identified MP particles in deposition samples, found in Kernchen et al. (2022)

understand sources, abundances, transportation patterns, and physicochemical processes of airborne synthetic polymers with aerodynamic diameter lower than 10 μm.

2 Conclusion

Plastic pollution is the most emerging environmental concern due to the specific nature of synthetic polymers, which under the influence of sunlight, temperature, mechanical effect, wind and wave action degrade into smaller fragments. Microplastics includes synthetic airborne particles with negative effects on ecosystems, marine biota, and humans. Although there has been a significant advancement in methodologies for microplastic analysis, the further method development is needed to obtain reliable quantitative information of MP in the size below 10 μm. For that purpose, analytical technique based on the Curie Point Pyrolysis GC MS is going to be implemented to determine the mass concentration of airborne MP. The developed method will be evaluated for its application to particulate matter samples, collected in urban and rural environments, with the goal to make a comparison between different locations. Furthermore, the contribution of long-term measurements of synthetic airborne particles would be interpreted in terms of the number and mass concentration of MP and by identification of potential markers for the specific sources of MP.

References

Barbosa, F., Adeyemi, J. A., Bocato, M. Z., Comas, A., & Campiglia, A. (2020). A critical viewpoint on current issues, limitations, and future research needs on micro- and nanoplastic studies: From the detection to the toxicological assessment. *Environmental Research, 182*, 109089.

Barnes, D. K. A., Galgani, F., Thompson, R. C., & Barlaz, M. (2009). Accumulation and fragmentation of plastic debris in global environments. *Philosophical Transactions of the Royal Society B: Biological Sciences, 364*(1526), 1985–1998.

Brahney, J., Mahowald, N., Prank, M., Cornwell, G., Klimont, Z., Matsui, H., & Prather, K. A. (2021). Constraining the atmospheric limb of the plastic cycle. *Proceedings of the National Academy of Sciences, 118*(16), e2020719118.

Cai, L., Wang, J., Peng, J., Wu, Z., & Tan, X. (2018). Observation of the degradation of three types of plastic pellets exposed to UV irradiation in three different environments. *Science of the Total Environment, 628–629*, 740–747.

Geyer, R., Jambeck, J. R., & Law, K. L. (2017). Production, use, and fate of all plastics ever made. *Science Advances, 3*(7), e1700782.

Ivleva, N. P. (2021). Chemical analysis of microplastics and nanoplastics: Challenges, advanced methods, and perspectives. *Chemical Reviews, 121*(19), 11886–11936.

Jambeck, J. R., Geyer, R., Wilcox, C., Siegler, T. R., Perryman, M., Andrady, A., Narayan, R., & Law, K. L. (2015). Plastic waste inputs from land into the ocean. *Science, 347*(6223), 768–771.

Kale, S. K., Deshmukh, A. G., Dudhare, M. S., & Patil, V. (2015). Microbial Degradation Of Plastic - A Review. *Journal of Biochemical Technology, 6*, 952–961.

Kernchen, S., Löder, M. G. J., Fischer, F., Fischer, D., Moses, S. R., Georgi, C., Nölscher, A. C., Held, A., & Laforsch, C. (2022). Airborne microplastic concentrations and deposition across the Weser River catchment. *Science of the Total Environment, 818*, 151812.

Lebreton, L., & Andrady, A. (2019). Future scenarios of global plastic waste generation and disposal. *Palgrave Communications, 5*(1), 6.

Leslie, H. A., & Depledge, M. H. (2020). Where is the evidence that human exposure to microplastics is safe? *Environmental International, 142*, 105807.

Leslie, H. A., van Velzen, M. J. M., Brandsma, S. H., Vethaak, A. D., Garcia-Vallejo, J. J., & Lamoree, M. H. (2022). Discovery and quantification of plastic particle pollution in human blood. *Environment International, 163*, 107199.

Liu, C., Li, J., Zhang, Y., Wang, L., Deng, J., Gao, Y., Yu, L., Zhang, J., & Sun, H. (2019). Widespread distribution of PET and PC microplastics in dust in urban China and their estimated human exposure. *Environmental International, 128*, 116–124.

Primpke, S., Christiansen, S., Cowger, W., De Frond, H., Deshpande, A., Fischer, M., Holland, E., Meyns, M., O'Donnell, B., Oßmann, B., Pittroff, M., Sarau, G., Scholz-Böttcher, B., & Wiggin, K. (2020). Critical assessment of analytical methods for the harmonized and cost-efficient analysis of microplastics. *Applied Spectroscopy, 74*(9), 1012–1047.

Rochman, C. M., Brookson, C., Bikker, J., Djuric, N., Earn, A., Bucci, K., Athey, S., Huntington, A., McIlwraith, H., Munno, K., De Frond, H., Kolomijeca, A., Erdle, L., Grbic, J., Bayoumi, M., Borrelle, S. B., Wu, T., Santoro, S., Werbowski, L. M., … Hung, C. (2019). Rethinking microplastics as a diverse contaminant suite. *Environmental Toxicology and Chemistry, 38*(4), 703–711.

Schwabl, P., Köppel, S., Königshofer, P., Bucsics, T., Trauner, M., Reiberger, T., & Liebmann, B. (2019). Detection of various microplastics in human stool. *Annals of Internal Medicine, 171*(7), 453–457.

Sebille, E., Spathi, C., & Gilbert, A. (2016). The ocean plastic pollution challenge: towards solutions in the UK.

Ugwu, K., Herrera, A., & Gómez, M. (2021). Microplastics in marine biota: A review. *Marine Pollution Bulletin, 169*, 112540.

Veerasingam, S., Ranjani, M., Venkatachalapathy, R., Bagaev, A., Mukhanov, V., Litvinyuk, D., Mugilarasan, M., Gurumoorthi, K., Guganathan, L., Aboobacker, V. M., & Vethamony, P. (2021). Contributions of Fourier transform infrared spectroscopy in microplastic pollution research: A review. *Critical Reviews in Environmental Science and Technology, 51*(22), 2681–2743.

Zhang, Y., Kang, S., Allen, S., Allen, D., Gao, T., & Sillanpää, M. (2020). Atmospheric microplastics: A review on current status and perspectives. *Earth-Science Reviews, 203*, 103118.

Zimmermann, L., Göttlich, S., Oehlmann, J., Wagner, M., & Völker, C. (2020). What are the drivers of microplastic toxicity? Comparing the toxicity of plastic chemicals and particles to Daphnia magna. *Environmental Pollution, 267*, 115392.

Determination of Ambient Air Quality Status Through Assessment of Particulate Matter and Selected Gases

George Oindo Achieng' and Dickson Mubera Andala

Abstract

Ambient air quality measurement is an integral part of environmental monitoring since human exposure to air pollutants could result in a range of health hazards depending on the type of pollutant, concentration, length and frequency of exposure and the accompanying toxicity of the specific pollutant. The objective of this survey was to measure the concentrations of particulate matter ($PM_{2.5}$ and PM_{10}), Nitrogen Oxides (NO_x), Ozone (O_3), Sulphur Dioxide (SO_2) and Carbon Monoxide (CO) from three selected locations A, B and C in Kenya using the portable HAZ-SCANNER system (HIM 6000). The sole choice of the three locations was informed by the nature of activities and humans working at them, thus, justifying their sensitivity. This endeavour would present some vital data on the ambience status with respect to the detected and quantified air pollutants that are potential for health risks. Minimum, maximum and average concentrations of particulate matter ($PM_{2.5}$ and PM_{10}), Nitrogen Oxides (NO_x), Ozone (O_3), Sulphur Dioxide (SO_2) and Carbon Monoxide (CO) were recorded with utmost precision. The results indicated that NO_x, O_3, $PM_{2.5}$ and PM_{10} in location A, $PM_{2.5}$ and PM_{10}, NO_x, SO_2 and O_3 in location B, and NO_x, $PM_{2.5}$ and PM_{10} in location C surpassed the permissible confines stipulated in the World Health Organization (WHO) air quality guidelines (AQG) of 2021. The presented results will form the basis from which the managements of the respective organizations will draw their safety measures hence offer sustainable protection to their workers through periodical monitoring and applicable personal protective equipment (PPE).

Keywords

Particulate matter · Gaseous pollutants · Haz-scanner system

1 Introduction

1.1 Background

Air pollution is defined as all negative effects of any sources that contribute to the contamination of the atmosphere and subsequent deterioration of the ecosystem. Air pollution is caused by both human intrusions and natural phenomena in which pollutants including materials in solid, liquid, and gas phases make part of (Vallero, 2007). Air quality evaluation is, therefore, imperative for gauging the nature of population exposure to air pollution. Assessment of population exposure is essential for health impact assessment, which in turn is fundamental for developing plans for air quality management and protecting the public health (Ghorani-Azam et al., 2016). This can be done through specific measurement methods that are subsequently considered for determining and attaining compliance based on the formulated National Ambient Air Quality Standard (Chow, 1995). Ray and Lahiri (2010) reported that chronic exposure to vehicular pollution reduces lung function, increases blood pressure, suppresses immunity, and enhances cancer risk in the lung. Cardiovascular diseases which comprise the rise in blood pressure or hypertension, ischemic heart disease, lung cancer, pneumonia, chronic obstructive pulmonary disease which are mainly caused by the exposure to PM, NO_x, CO and O_3 (Sander et al., 2015). Wood burning, soil dust, brick kilns, fugitive Pb, road dust,

G. O. Achieng' (✉)
Maseno University, P.O. Box 333-40105, Maseno, Kenya
e-mail: georgeoindoachieng@gmail.com

D. M. Andala
Department of Chemistry, Multimedia University of Kenya, P.O. Box 15653–00503, Nairobi, Kenya
e-mail: dandala@mmu.ac.ke

K. W. Fomba et al. (eds.), *Advances in Air Quality Research in Africa*,
Advances in Science, Technology & Innovation, https://doi.org/10.1007/978-3-031-53525-3_11

Zn sources, motor vehicles, and sea salt were identified as the main sources of air pollution with $PM_{2.5}$ in Bangladesh (Begum & Hopke, 2019). In this regard, the Government of Bangladesh began considering various interventions to reduce the emissions from these sources by promoting the replacement of diesel/petrol automobiles with compressed natural gas (CNG) vehicles, increasing traffic speed in the city, and introducing green technologies for brick production. The World Health Organization (WHO) considers particle pollution, ground-level Ozone (O_3), Carbon Monoxide (CO), sulphur oxides (SOx), nitrogen oxides (NOx), and lead (Pb) as the six major air pollutants that harm both human health and the ecosystem. The pollutants are suspended materials such as dust, fumes, smokes, mists, gaseous pollutants etc. that result into vulnerability to many diseases including different types of cancers (Loomis et al., 2014). The objective of this research was to measure the concentrations of particulate matter ($PM_{2.5}$ and PM_{10}), Nitrogen Oxides (NO_x), Ozone (O_3), Sulphur Dioxide (SO_2) and Carbon Monoxide (CO) from three locations A, B and C in Kenya, and ascertain their compliance with the WHO requirements based on the Air Quality Guideline (AQG).

Table 1 The levels of pollutants measured in the air

Location	Parameter	Minimum	Maximum	Average	AQG
A	NO_x (ppb)	55.60	113.40	**83.84**	**12.00**
	O_3 (ppb)	50.58	55.40	**53.81**	**47.00**
	SO_2 (ppb)	<0.001	3.28	0.80	**14.00**
	CO (ppm)	1.54	2.49	1.74	**2.00**
	$PM_{2.5}$ (µg/m³)	18.24	36.84	**32.33**	**15.00**
	PM_{10} (µg/m³)	274.90	286.60	**284.09**	**45.00**
B	NO_x (ppb)	10.91	229.46	**111.43**	**12.00**
	O_3 (ppb)	14.12	127.70	**60.63**	**47.00**
	SO_2 (ppb)	18.50	44.57	**32.11**	**14.00**
	CO (ppm)	<0.001	0.317	0.144	**2.00**
	$PM_{2.5}$ (µg/m³)	1.00	16.03	7.54	**15.00**
	PM_{10} (µg/m³)	2.00	1013.00	**375.77**	**45.00**
C	NO_x (ppb)	1.00	60.25	**48.40**	**12.00**
	O_3 (ppb)	<0.001	52.23	41.78	**47.00**
	SO_2 (ppb)	<0.001	<0.001	<0.001	**14.00**
	CO (ppm)	<0.001	<0.001	<0.001	**2.00**
	$PM_{2.5}$ (µg/m³)	<0.001	165.50	**132.42**	**15.00**
	PM_{10} (µg/m³)	20.00	167.30	**137.87**	**45.00**

N/B: The Standard Limits are quoted from the WHO air quality guidelines (AQG) of 2021

2 Methodology

The HAZ-SCANNER system (HIM 6000) was first switched on and allowed to warm up for 15 minutes during which the logging function was disabled. The system was set to sample and log data after every ten minutes for four hours. The minimum, maximum and average concentrations of particulate matter ($PM_{2.5}$ and PM_{10}), Nitrogen Oxides (NO_x), Ozone (O_3), Sulphur Dioxide (SO_2) and Carbon Monoxide (CO) were recorded from locations A, B and C in Kenya which were situated at a health facility, oil storage and transporting company, and road-side next to a tarmacked road, correspondingly. The exclusive choice of the three locations was informed by the nature of the activities and humans working at them, thus, justifying their sensitivity. The levels of the pollutants were evaluated against the World Health Organization (WHO) air quality guidelines of 2021 (Zhao et al., 2022). Note that NO_2 AQG level was used for NO_x in drawing conclusion on compliance status.

3 Results and Discussion

The levels of particulate matter ($PM_{2.5}$ and PM_{10}), Nitrogen Oxides (NO_x), Ozone (O_3), Sulphur Dioxide (SO_2) and Carbon Monoxide (CO) recorded from locations A, B and C are given in Table 1.

The sampling location A posted SO_2 and CO levels that were compliant with the WHO air quality guidelines of 2021. On the other hand, the levels of $PM_{2.5}$, PM_{10}, NO_x, SO_2 and O_3 in location B were non-compliant. In addition, the concentrations of all the parameters except NO_x, $PM_{2.5}$ and PM_{10} assessed in location C achieved the WHO compliance confines. In a review by Hůnová (2020) on the past and present Ambient Air Quality in the Czech Republic, the levels of O_3 vastly exceeded the standard limits in spite of a significant reduction in overall emissions. In another research, Awokola et al. (2020) analysed one-year longitudinal $PM_{2.5}$ measurement in thirteen sites and reported average levels between 0.5 and 91.1 µg/m³. A study by Chen et al. (2007) reported that exposure to SO_2 can cause damages to the eyes, mucous membranes, the skin, and respiratory tracts. In addition, the researchers established that Bronchospasm, pulmonary edema, pneumonitis, and acute airway obstruction are the most common clinical findings associated with exposure to SO_2. Reports have shown that the most common complication of NO_x toxicity is coughing and wheezing, however, the eyes, nose or throat irritations, headache, dyspnoea, chest pain, diaphoresis, fever, bronchospasm, and pulmonary edema may also occur (Ghorani-Azam et al., 2016; Hesterberg et al., 2009). Moreover, Sadeghi et al. (2015) established that PM pollution is linked with most of pulmonary and cardiac-associated morbidity and mortality. The exposure to PM, NO_x, CO and O_3 have been associated with the rise in blood pressure or hypertension, ischemic heart disease, lung cancer, pneumonia, and chronic obstructive pulmonary disease (Sander et al., 2015). The results that indicate non-compliance to the WHO standard limit requirements draw attention to the managements of the respective organizations to reflect on the safety measures in order to offer sustainable protection to their workers through periodical monitoring and applicable personal protective equipment.

4 Conclusion

Considering the World Health Organization (WHO) air quality guidelines of 2021, NO_x, O_3, $PM_{2.5}$ and PM_{10} in location A, $PM_{2.5}$ and PM_{10}, NO_x, SO_2 and O_3 in location B, and $PM_{2.5}$ and PM_{10} in location C surpassed the permissible confines, thus, rendering the three sampling locations polluted with these parameters. The results as so presented will form the basis from which the managements of the respective organizations will draw their safety measures hence offer sustainable protection to their workers through periodical monitoring and applicable personal protective equipment.

Acknowledgements The authors of this work sincerely acknowledge CSI International Limited quality testing laboratory for the sponsorship.

References

Awokola, B. I., Okello, G., Mortimer, K. J., Jewell, C. P., Erhart, A., & Semple, S. (2020). Measuring air quality for advocacy in Africa (MA3): feasibility and practicality of longitudinal ambient $PM_{2.5}$ measurement using low-cost sensors. *International Journal of Environmental Research and Public Health, 17*(19), 7243.

Begum, B. A., & Hopke, P. K. (2019). Identification of sources from chemical characterization of fine particulate matter and assessment of ambient air quality in Dhaka. *Bangladesh. Aerosol and Air Quality Research, 19*(1), 118–128.

Chow, J. C. (1995). Measurement methods to determine compliance with ambient air quality standards for suspended particles. *Journal of the Air & Waste Management Association, 45*(5), 320–382.

Ghorani-Azam, A., Riahi-Zanjani, B., & Balali-Mood, M. (2016). Effects of air pollution on human health and practical measures for prevention in Iran. *Journal of Research in Medical Sciences: The Official Journal of Isfahan University of Medical Sciences, 21*, 1–65.

Hesterberg, T. W., Bunn, W. B., McClellan, R. O., Hamade, A. K., Long, C. M., & Valberg, P. A. (2009). Critical review of the human data on short-term nitrogen dioxide (NO_2) exposures: Evidence for NO_2 no-effect levels. *Critical Reviews in Toxicology, 39*(9), 743–781.

Hůnová, I. (2020). Ambient air quality in the Czech Republic: Past and present. *Atmosphere, 11*(2), 214.

Chen, T. M., Gokhale J., Shofer, S., & Kuschner, W. G. (2007). Outdoor air pollution: nitrogen dioxide, sulphur dioxide, and carbon monoxide health effects. *The American Journal of the Medical Sciences, 333*(4), 249–256.

Loomis, D., Huang, W., & Chen, G. (2014). The International Agency for Research on Cancer (IARC) evaluation of the carcinogenicity of outdoor air pollution: Focus on China. *Chinese Journal of Cancer, 33*(4), 189–196.

Ray, M. R., & Lahiri, T. (2010). *Air pollution and its effects on health–Case studies* (p. 11). Pp.

Sadeghi, M., Ahmadi, A., Baradaran, A., Masoudipoor, N., & Frouzandeh, S. (2015). Modeling of the relationship between the environmental air pollution, clinical risk factors, and hospital mortality due to myocardial infarction in Isfahan Iran. *Journal of Research in Medical Sciences: The Official Journal of Isfahan University of Medical Sciences, 20*(8), 757.

Sander, K., Mira-Salama, D., & Feuerbacher, A. (2015). *The cost of air pollution*. Pp 50.

Vallero, D. (2007). *Fundamentals of air pollution*. Academic press.

Zhao, B., Shi, S., & Ji, J. S. (2022). The WHO Air Quality Guidelines 2021 promote great challenge for indoor air. *Science of the Total Environment, 827*, 154–376.

Data, Modelling and Forecasting

Determination of the Dispersion of Carbon Monoxide, Sulphur Dioxide and Nitrogen Oxides from Selected Production Plants Along Thika Road and Mombasa Road in Nairobi County Kenya

Jermaine O. Omulami, Patrick K. Tum, Rachael E. N. Njogu, Dickson M. Andala and George Oindo Achieng'

Abstract

Uncontrolled air pollution leads to climate change and contributes significantly to the health burden thus slowing down economic growth. In this paper, air-dispersion modelling has been studied as a viable technique to effectively predict the contribution of specific emission sources to the ambient air quality from two Production Plants (A and B) situated within an urban setting. The findings indicate the mean maximum 1-h concentration of CO was 41.87 mg/m^3 and 0.00233 mg/m^3, the 24-h concentration of SO_2 was 5.45 µg/m^3 and 65.22 µg/m^{3}and 24-h maximum concentration of NO_x was 31.38 µg/m^3 and 10.93 µg/m^3 for Production Plants A and B, respectively. In plant A, the mean maximum concentration of CO exceeded the limits set by both the National Environmental Management Authority (NEMA Kenya) and the World Health Organization (WHO) and those of NO_x exceeded limits set by WHO. The levels of CO and NO_x emitted from plant B were within tolerable limits; however, the levels of SO_2 exceeded the limits set by WHO.

Keywords

AERMOD · Air dispersion modelling · Air pollution · Air quality forecasting

1 Introduction

1.1 Background

The industrial revolution has not only improved the economies of various countries but also the quality of life through employment (Dogaru, 2020; Lucas, 2004). However, various studies have shown that the use of fossil fuel and biomass is accompanied by the release of harmful emissions such as NO_x, SO_2, CO, CO_2, PM_{10}, $PM_{2.5}$ and hydrocarbons that cause air pollution into the atmosphere (Ali et al., 2011). Power-generating plants using coal are the main source of green house gases (GHG) such as CO_2 and CH_4, NO_x, SO_2, Hg, and $PM_{2.5}$ (Sahu et al., 2021). In the USA, electricity generation is one of the largest contributors of $PM_{2.5}$ in the air with the lower-income areas having higher exposures (Thind et al., 2019). Air pollution is linked to global warming, respiratory infections and increased mortality rates (Croft et al., 2019; Jiang et al., 2016). Due to the aforementioned concerns, some of the techniques developed to control air pollution include the utilization of renewable energy sources, emission control through absorption, adsorption, condensation and incineration (Schnelle & Brown, 2001) and the application of various air quality forecasting techniques such as air dispersion modelling (CENR, 2001). Air Dispersion modelling tools such as the American Meteorological Society/Environmental Protection Agency Regulatory Model, AERMOD, can be used to carry out air quality impact assessments from various pollution sources (Schnelle & Brown, 2001).

AERMOD is a Gaussian plume modelling system (Visscher, 2013) based on the assumption emission released

J. O. Omulami · P. K. Tum · R. E. N. Njogu
University of Nairobi, P.O. Box 30197-00100, Nairobi, Kenya
e-mail: jermaine.omulami@gmail.com

P. K. Tum
e-mail: patricktum@uonbi.ac.ke

R. E. N. Njogu
e-mail: njogurachael@uonbi.ac.ke

D. M. Andala
Multimedia University of Kenya, P.O. Box 15653-00503, Nairobi, Kenya
e-mail: dandala@mmu.ac.ke

G. O. Achieng' (✉)
Maseno University, P.O. Box 333-40105, Maseno, Kenya
e-mail: georgeoindoachieng@gmail.com

K. W. Fomba et al. (eds.), *Advances in Air Quality Research in Africa*,
Advances in Science, Technology & Innovation, https://doi.org/10.1007/978-3-031-53525-3_12

at a given rate, travelling along the x, y and z-axis to assume a Gaussian concentration profile (Brusca et al., 2016; Visscher, 2013). AERMOD is made of a meteorological (AERMET) and a terrain (AERMAP) pre-processor (Jittra et al., 2015). AERMET is a meteorological data pre-processor used to calculate planetary boundary parameters (Zade & Ingole, 2015).

2 Research Question/Problem

The State of Global Air 2019 reported up to 19,000 premature deaths due to air pollution in Kenya (deSouza, 2020). In 2013, the residents of Owino Uhuru settlement near Mombasa reported lead poisoning which was linked to the lead battery recycling company in the area (Mwanza, 2021). To prevent future air pollution effects such as the above cases, this study investigated the dispersion of various pollutants; CO, SO_2 and NO_x from selected Production Plants, A and B.

3 Methodology

3.1 Study Area

Two factories, along Mombasa Road (A) and Thika Road (B) in Kenya, were identified for this study. Plants A and B utilised firewood and heavy fuel oil (HFO) as sources of energy, respectively. Plant A is surrounded by low to middle-income residential areas including Allsops, Buruburu, Baba Dogo, Kariobangi and Lucky Summer on the Eastern side of the facility. Plant B is also surrounded by low to middle-income residential areas such as Cabanas, Makadara, Donholm and Syokimau.

3.2 Air Dispersion Model

AERMOD is a Gaussian plume steady-state model recommended by the USEPA for the investigation of the effects of emissions on receptors within a 50 km radius from the point of emissions. AERMOD input data included terrain data, land cover data, meteorological data, source location and source emission data.

3.3 Data Input

Source Data: Source location data and emission concentrations of CO, SO_2 and NO_x data for each plant were obtained from NEMA Kenya. The source data included the stack emission UTM Coordinates, stack height and diameter, stack exit temperature and exit volumetric and emission flow rate. For compliance purposes, manufacturing industries in Kenya are required to submit emission reports on quarterly basis. The data obtained were those reported for the four quarters of 2020.

Meteorological Data: Hourly surface meteorological data for the years 2018 to 2020 was obtained from NCEI as reported by Wilson Airport and Dagoretti meteorological weather station. The data contained hourly surface pressure, cloud cover, wind direction and speed, relative humidity and opaque sky cover. The upper air sounding data for the years 2018 to 2020 was obtained from NOAA/ESRL as reported by Dagoretti meteorological weather station. Meteorological data was processed using the AERMET pre-processor. AERMET also contain the AERSURFACE used to process the land cover surface use data.

Terrain Data: DEM Terrain data, collected during the SRTM of the area surrounding the plants were obtained from the USGS. Terrain data was processed using AERMAP pre-processor.

3.4 Output

The simulation was done to obtain concentration for the 1st highest 1-h period for CO and the 24-h period for SO_2 and NO_x. The site domain used contained a total of 10,000 uniformly spaced receptors each with a grid resolution of 100 m by 100 m. The concentrations were calculated at a flagpole height of 1.5 m, the approximate breathable height. The dispersion results obtained were presented as plot files. The mean maximum concentrations for each of the four quarters (Q1, Q2, Q3 and Q4) were compared to the ambient air quality limits as set by NEMA and WHO.

Table 1 Mean maximum concentration

Maximum Concentration							Ambient tolerance limit	
		Q1	Q2	Q3	Q4	Mean	NEMA	WHO
CO (1-h) mg/m^3	A	52.69	45.21	47.32	33.07	41.87	4	4
	B	0.0038	0.0046	0.0015	0.00097	0.0023		
SO_2 (24-h) µg/m^3	A	7.23	6.06	5.29	3.22	5.45	80	40
	B	103.0	66.4	49.884	41.6	65.22		
NO_x(24-h) µg/m^3	A	33.0	32.0	37.2	23.30	31.38	80	25
	B	15.80	13.80	5.72	8.40	10.93		

Source of tolerance limit (NEMA, 2014; WHO, 2021)

4 Results and Discussion

4.1 Mean Maximum Concentration

The maximum concentrations of CO, SO_2 and NO_x were at a downwind distance of 71.72 m heading 225.28^0 and 84.02 m heading 249.15^0 from A and B respectively. The average mean maximum concentrations at 1.5 m above the ground (see Table 1).

The average 1-h maximum concentrations of CO were 41.87 mg/m^3 and 0.0023 mg/m^3 for A and B respectively. CO emissions for Plant A exceeded the tolerable limits set by both WHO and NEMA while that released by Plant B was lower.

The average 24-h maximum concentrations of SO_2 were 5.45 µg/m^3 and 65.22 µg/m^3 for Plant A and B respectively. SO_2 emissions for Plant A were below tolerable limits set by both WHO and NEMA while that released by Plant B was below the limit set by NEMA but higher than that set by WHO.

The average 24-h maximum concentrations of NO_x were 31.38 µg/m^3 and 10.93 µg/m^3 for A and B respectively. NO_x emission for Plant A was below tolerable limits set by NEMA but exceeded the WHO limits while that released by Plant B was below the limit set by both NEMA and WHO.

4.2 Dispersion (Plot Files)

From Plant A, CO is evenly distributed in all directions, with most emissions moving in the North, West and South directions, NO_x and SO_2 predominantly move in the North of the stationary point. From Plant B, the emissions predominantly move to the western side of the facility (see Fig. 1).

5 Conclusion

Plant A contributed to a higher ambient concentration of CO and NO_x while Plant B contributed to a higher SO_2. The high concentration of CO and NO_x in Plant A is likely to be attributed to the use of biomass as their source of energy while the high concentration of SO_2 in B is due to the use of heavy fuel oil (HFO). The concentrations can be reduced by the use of alternative fuels such as electricity or the installation of air pollution control technology that targets the above gases. CO and NO_x from Plant A can be reduced by checking the air-to-fuel ratio to ensure complete combustion and subsequent use of NaOH-based scrubbers (Schnelle & Brown, 2001). In addition, increasing the stack height would reduce the maximum predicted concentration by reducing the downwash effect (Schnelle & Brown, 2001; Schulman & Scire, 2012).

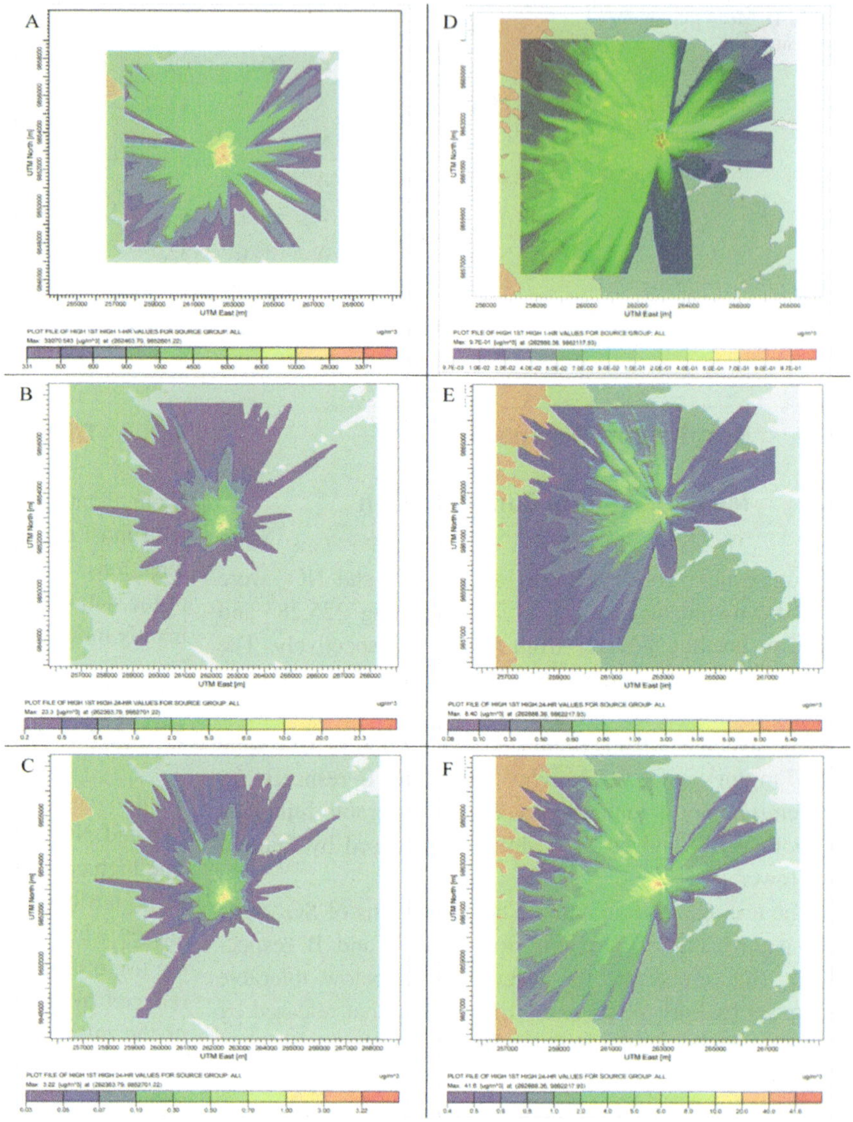

Fig. 1 Plot file for **a** 1-h CO **b** 24-h NO_x **c** 24-h SO_2 Facility A and **a** 1-h CO **b** 24-h NO_x **c** 24-h SO_2 Facility B

Acknowledgements We would like to acknowledge CSI International Limited Laboratory for their assistance and sponsorship in this work.

References

Ali, M. B., Saidur, R., & Hossain, M. S. (2011). A review on emission analysis in cement industries. *Renewable and Sustainable Energy Reviews, 15*(5), 2252–2261.

Brusca, S., Famoso, F., Lanzafame, R., Mauro, S., Garrano, A. M. C., & Monforte, P. (2016). Theoretical and experimental study of gaussian plume model in small scale system. *Energy Procedia, 101*, 58–65.

CENR. (2001). Air quality forecasting: A review of federal programs and research needs air quality research subcommittee of the committee on environment and natural resources, 1–26.

Croft, D. P., Zhang, W., Lin, S., Thurston, S. W., Hopke, P. K., Masiol, M., Squizzato, S., van Wijngaarden, E., Utell, M. J., & Rich, D. Q. (2019). The association between respiratory infection and air pollution in the setting of air quality policy and economic change. *Annals of the American Thoracic Society, 16*(3), 321–330.

deSouza, P. (2020). Air pollution in Kenya: A review. *Air Quality, Atmosphere Health, 13*(12), 1487–1495.

Dogaru, L. (2020). The main goals of the fourth industrial revolution. *Renewable Energy Perspectives. Procedia Manufacturing, 46*, 397–401.

Jiang, X. Q., Mei, X. D., & Feng, D. (2016). Air pollution and chronic airway diseases: what should people know and do? *Journal of Thoracic Disease, 8*(1).

Jittra, N., Pinthong, N., & Thepanondh, S. (2015). Performance evaluation of AERMOD and CALPUFF air dispersion models in industrial complex area. *Air, Soil and Water Research*, 87–95.

Lucas, R. E. (2004). Economic education bulletin. In *The industrial revolution: Past and future*. www.aier.org

Mwanza, R. (2021). Toxic Spaces, Community voices, and the promise of environmental human rights: Lessons on the owino uhuru pollution incident in Kenya. *Nordic Journal of Human Rights, 38*(4), 279–304.

NEMA. (2014). Legal notice no. 34 e environmental management and co-ordination act (no 8 of 1999) the environmental management and co-ordination (air quality) regulations, 2014.

Sahu, S. K., Zhu, S., Guo, H., Chen, K., Liu, S., Xing, J., Kota, S. H., & Zhang, H. (2021). Contributions of power generation to air pollution and associated health risks in India: Current status and control scenarios. *Journal of Cleaner Production, 288*, 125587.

Schnelle, J. K. B., & Brown, C. A. (2001). *Air pollution control technology handbook* (1st ed.). CRC Press.

Schulman, L. L., & Scire, J. S. (2012). Building downwash modeling with AERMOD. In *10th Conference on Air Quality Modeling: RTP, NC.*

Thind, M. P. S., Tessum, C. W., Azevedo, I. L., & Marshall, J. D. (2019). Fine particulate air pollution from electricity generation in the US: Health impacts by race, income, and geography. *Environmental Science and Technology, 53*(23), 14010–14019.

Visscher, A. (2013). De.: aermod and aermet: A detailed description. In *Air dispersion modeling a foundation and application* (pp. 491–513).

WHO. (2021). WHO global air quality guidelines: particulate matter ($PM_{2.5}$ and PM_{10}), ozone, nitrogen dioxide, sulfur dioxide and carbon monoxide.

Zade, S., & Ingole, N. W. (2015). Air Dispersion modelling to assess ambient air quality impact due to carbon industry. *International Journal of Research Studies in Science, Engineering and Technology, 2*(7), 53.

Forecasting with the GEOS-CF System and Other NASA Resources to Support Air Quality Management

Carl Malings, K. Emma Knowland, Christoph Keller, Stephen Cohn, Bryan Duncan and Nathan Pavlovic

Abstract

Air quality (AQ) is a major and growing concern for public health around the world. Economic development, population growth, and climate change are all expected to exacerbate already poor AQ in many regions. Furthermore, AQ is often only sparsely monitored with reference-grade in-situ instruments. NASA resources and products have the potential to help in addressing this AQ data gap. The GEOS-CF (Goddard Earth Observing System—Composition Forecasting) global atmospheric composition modeling system is run each day at a global scale to provide recent estimates and five-day forecasts at hourly temporal resolution of atmospheric constituents relevant to AQ. NASA satellite missions (along with those of other space agencies) provide remotely sensed estimates of atmospheric composition relevant to AQ. This paper gives a brief overview of these capabilities, and outlines the efforts underway to combine model forecasts, satellite retrievals, and surface-based measurements to provide more comprehensive and accurate estimates and forecasts of local AQ which will be broadly applicable and accessible globally.

Keywords

Forecasting · NASA · Models · Satellites · GEOS-CF · GEOS-Chem

C. Malings · C. Keller
Morgan State University, Baltimore, MD 21251, USA

C. Malings (✉) · K. E. Knowland · C. Keller · S. Cohn · B. Duncan
NASA Goddard Space Flight Center, Greenbelt, MD 20771, USA
e-mail: carl.a.malings@nasa.gov

N. Pavlovic
Sonoma Technology, Inc., Petaluma, CA 94954, USA

1 Introduction

Poor air quality (AQ) is a major and growing global public health concern (Babatola, 2018). It is estimated that the Global South is affected disproportionately by poor AQ, although this is difficult to verify empirically (Katoto et al., 2019). Ideally, AQ can be tracked by a national regulatory reference-grade monitor (RGM) network (Snyder et al., 2013). Such networks are in place in several counties, many of which make their AQ data freely available to improve transparency (OpenAQ Homepage, 2023). Many areas, however, including most African nations, have not yet established such networks (or not made their data publicly available), in part due to the high purchase and operation costs of RGM (Amegah & Agyei-Mensah, 2017). Low-cost AQ sensors (LCS) can help address this data gap, having already increased surface AQ monitoring density by up to an order of magnitude in several regions (see Fig. 1). LCS require localized calibration to account for errors and biases, which can be difficult in the absence of local RGM (Amegah et al., 2021; Bittner et al., 2022; Giordano et al., 2021). Nevertheless, use of LCS in Africa is an area of active research (Subramanian et al., 2020; Bainomugisha et al., 2023; McFarlane et al., 2021a, 2021b; Raheja et al., 2022).

Other sources are also available to help address the African AQ data need. Remote sensing from satellite instruments provides regular retrievals of atmospheric properties and composition relevant to AQ. It should be noted that these often refer to column-integrated quantities, from which surface-level concentrations can be inferred. Global atmospheric composition models, such as the NASA GEOS-CF (Goddard Earth Observing System—Composition Forecasting) system, provide daily estimates and forecasts of AQ information (Keller et al., 2021a). Combining AQ information from models, satellites, and surface-based monitors (both RGM and LCS) can potentially help to address the as-yet unmet AQ data needs of

K. W. Fomba et al. (eds.), *Advances in Air Quality Research in Africa*,
Advances in Science, Technology & Innovation, https://doi.org/10.1007/978-3-031-53525-3_13

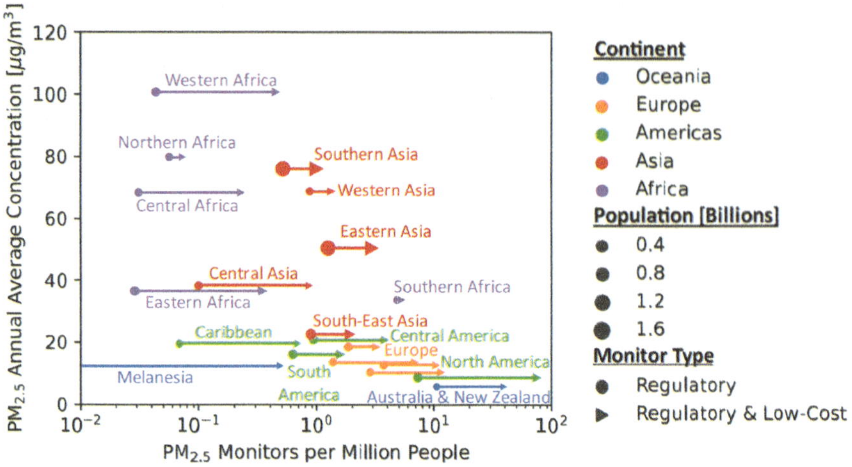

Fig. 1 There is an inverse tendency between poor AQ (annual average $PM_{2.5}$ concentration, vertical axis) and monitor density (horizontal axis, log scale). Supplementing RGM (circles) with LCS (triangles) helps to mitigate data gaps. Figure adapted from Malings et al. (2020) under the CC BY 4.0 license with additional information on RGM and LCS densities from the OpenAQ Homepage (2023)

many countries in Africa and the Global South generally (Duncan et al., 2021).

This chapter discusses (1) how the NASA GEOS-CF modeling system can support AQ estimation and forecasting in data-sparse regions, (2) how these forecasts can be combined with other data sources, and (3) how these tools can be made more accessible and useable by AQ managers in Africa.

2 Methods

2.1 The NASA GEOS-CF Modeling System

Each day, the NASA GEOS-CF modeling system produces global one-day replays and five-day forecasts of many atmospheric constituents relevant to AQ. Its output spatial resolution is 0.25° (approximately 25 km) with high temporal frequency (every 15 min and hourly-average). The online FLUID system (FLUID, 2023) allows air quality managers to easily visualize GEOS-CF outputs (see Fig. 2). A detailed overview and performance evaluation of GEOS-CF are provided by Keller et al. (2021a). In Africa, GEOS-CF generally reproduces seasonal trends in Ozone, NO_2, CO, and $PM_{2.5}$ observed by RGM. However, GEOS-CF underestimated Ozone in Africa during April through September by about 5 ppbv, and underestimated afternoon Ozone peaks by about 10 ppbv. Surface SO_2 is overestimated by GEOS-CF worldwide due to an outdated emission inventory. Notably, lack of RGM limited GEOS-CF validation in Africa, e.g., only nine sites contributed to $PM_{2.5}$ validation. This emphasizes the need for more surface AQ data in Africa.

2.2 Integrating Model, Satellite, and In-Situ Data

Model AQ forecasts such as those from GEOS-CF may be inaccurate and/or biased due to a combination of incomplete or outdated emissions information and the scale representativeness error between the model and specific locations within model grid cells (e.g., the sites of RGM). These biases can be adjusted for by combining multiple sources of information. For example, Bi et al. (2022) and Keller et al. (2021b) use machine learning techniques to adjust GEOS-CF outputs to better match historical RGM data at sites around the world, and apply these same techniques

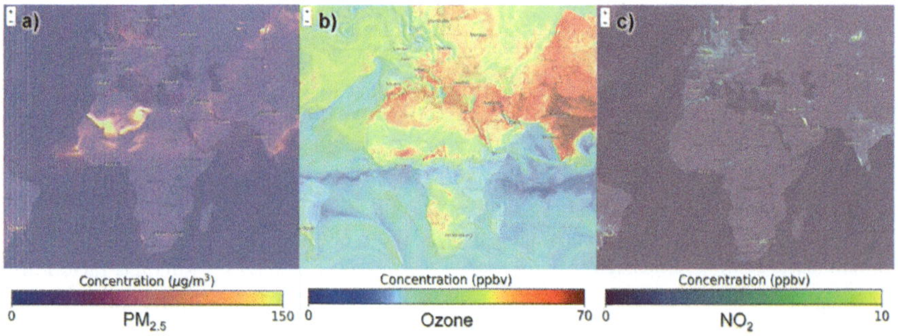

Fig. 2 Example of the online tool for visualization of GEOS-CF forecasts (FLUID, 2023). The plotted data represents surface $PM_{2.5}$ (**a**), ozone (**b**), and NO_2 (**c**) concentrations for 1130Z on May 9, 2023, from a forecast generated five days in advance

to bias-correct GEOS-CF forecasts at these RGM sites, improving forecast accuracy and reducing bias. A methodology for combining the GEOS-CF modeling system, TROPOMI satellite, and surface RGM data to improve NO_2 forecast accuracy in urban areas (see Fig. 3) is presented by Malings et al. (2021). Efforts are underway in collaboration with the World Resources Institute (World Resources Institute, 2023) and Sonoma Technology, Inc. (Sonoma Technology, 2023) to improve and expand these methods, making them more widely applicable and easily accessible by AQ managers around the world.

2.3 Data Accessibility and Capacity Building

NASA is engaged in numerous partnerships to make its AQ data and products accessible more broadly. To reduce the burden on end-users of downloading and processing large model and satellite data files locally, many of these datasets are being migrated to cloud computing platforms through the NASA Open-Source Science Initiative (2023). Additionally, a recent agreement between NASA and Google has led to the hosting of NASA data products (including GEOS-CF) in the Google Earth Engine (GEE) platform (2023). GEE will simplify the integration of GEOS-CF outputs with other AQ datasets to perform cloud-based data analysis (e.g., see Fig. 4) and to implement the AQ data integration system outlined by Malings et al. (2021) on a more globally applicable and accessible scale.

NASA also supports the development of AQ applications and capacity building globally. The NASA Health and Air Quality Applied Sciences Team (HAQAST) works in partnership with public health and AQ agencies across the globe to use NASA data and tools for the public benefit (NASA Health & Air Quality Applied Sciences Team, 2023). Lessons about the capabilities and use of NASA data products for AQ applications are provided by the Applied Remote Sensing Training (ARSET) program (NASA Applied Sciences ARSET Health & Air Quality Trainings, 2023). See Duncan et al. (2021) for more information about these resources.

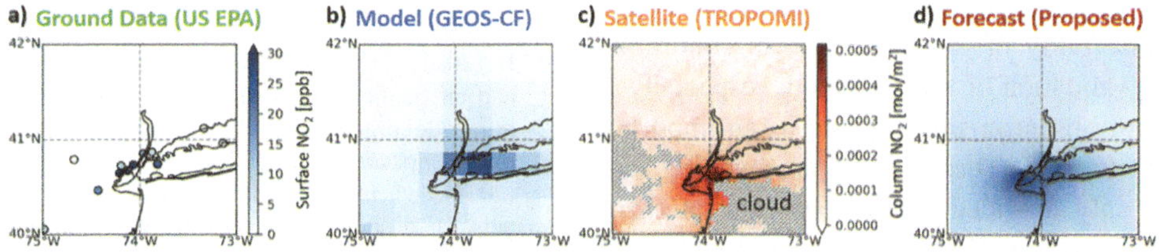

Fig. 3 AQ information from ground (**a**), model (**b**), and satellite (**c**) sources have been combined to improve sub-city-scale AQ forecasts (**d**) beyond what is available using each data source alone. Figure re-formatted and adapted from Malings et al. (2021) under the CC BY-NC 4.0 license

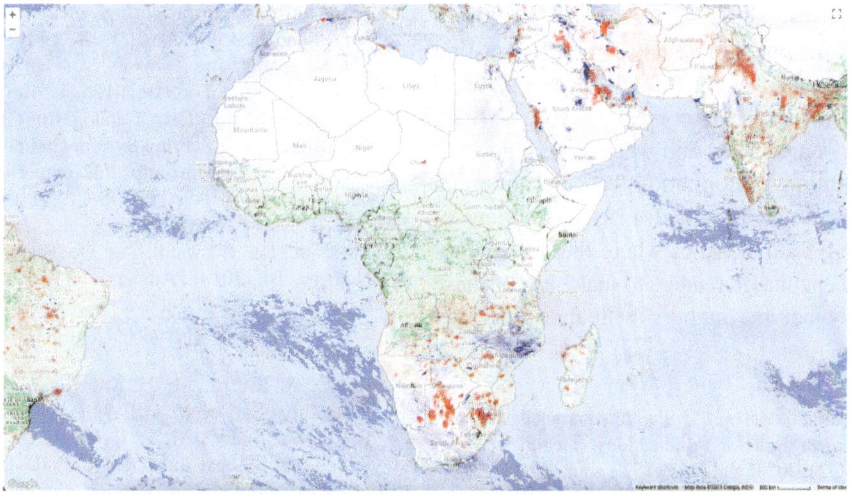

Fig. 4 A comparison between tropospheric column NO_2 concentrations from GEOS-CF and the TROPOMI satellite instrument (European Space Agency, 2021) performed in GEE for the period of 11–14 October 2022. Although the datasets are in general agreement across Africa, note the under-estimation by GEOS-CF over Mozambique (blue) and the over-estimations over South Africa (red). This demonstrates the use of GEE for easily intercomparing and combining AQ datasets from different sources and with different temporal and spatial resolutions

3 Results and Discussion

The NASA GEOS-CF modeling system can be an important tool for AQ estimation and forecasting. It has global coverage, relatively high spatial and temporal resolution (compared with similar global systems), and web-based resources exist to provide remote access to and interpretation of its products. However, there are certain inherent limitations of relying on model-based forecasts alone, including the use of potentially incomplete or outdated emissions inventories to inform the model and the potential for representativeness error between the model scale (representing national or regional conditions) and the scale of interest (e.g., neighborhood-scale AQ). Nevertheless, and especially in the absence of local AQ information from RGM or well-calibrated LCS, GEOS-CF can provide a valuable source of information and a forecasting capability. GEOS-CF is also useful for identifying and providing early warning for regionally transported pollutants such as smoke or dust.

Efforts have been made to integrate this model with other information sources, including satellite remote sensing information, RGM, and LCS, to reduce biases and improve the applicability of forecasts to local conditions. Furthermore, the validation of these methods, especially in areas with spare or absent in-situ AQ data, is an important topic of research interest. These efforts are ongoing, and this chapter has only provided a brief overview of them; more complete and up-to-date information can be found in the various papers and websites cited above.

4 Conclusions

There exist numerous freely available NASA resources which can be used to support AQ analysis and forecasting around the world (Duncan et al., 2021). These resources can play an important role in the context of Africa, where in-situ RGM networks are sparse and where emerging LCS networks require additional information for calibration and verification. This paper has provided a brief overview of these resources, along with websites where these resources can be accessed. We encourage readers to make use of these resources and to contact the authors with questions and feedback about them.

Acknowledgements The authors would like to acknowledge NASA funding through the Earth Science Applications: Health and Air Quality (grant # 21-HAQ21-0024) and HAQAST (grant # 20-HAQ20-0011) projects, as well as support from Google through the NASA and Google Earth Engine public-private partnership.

References

Amegah, A. K., & Agyei-Mensah, S. (2017). Urban air pollution in Sub-Saharan Africa: Time for action. *Environmental Pollution., 220*, 738–743. https://doi.org/10.1016/j.envpol.2016.09.042

Amegah, A. K., Dakuu, G., Mudu, P., & Jaakkola, J. J. K. (2021). Particulate matter pollution at traffic hotspots of Accra, Ghana: Levels, exposure experiences of street traders, and associated respiratory and cardiovascular symptoms. *Journal of Exposure Science & Environmental Epidemiology*. https://doi.org/10.1038/s41370-021-00357-x

Babatola, S. S. (2018). Global burden of diseases attributable to air pollution. *Journal of Public Health in Africa, 9*. https://doi.org/10.4081/jphia.2018.813.

Bainomugisha, E., Ssematimba, J., & Okure, D. (2023). Design considerations for a distributed low-cost air quality sensing system for urban environments in low-resource settings. *Atmosphere, 14*, 354. https://doi.org/10.3390/atmos14020354

Bi, J., Knowland, K. E., Keller, C. A., & Liu, Y. (2022). Combining machine learning and numerical simulation for high-resolution PM $_{2.5}$ concentration forecast. *Environmental Science & Technology, 56*, 1544–1556. https://doi.org/10.1021/acs.est.1c05578.

Bittner, A. S., Cross, E. S., Hagan, D. H., Malings, C., Lipsky, E., & Grieshop, A. P. (2022). Performance characterization of low-cost air quality sensors for off-grid deployment in rural Malawi. *Atmos. Meas. Tech., 15*, 3353–3376. https://doi.org/10.5194/amt-15-3353-2022

Duncan, B. N., Malings, C. A., Knowland, K. E., Anderson, D. C., Prados, A. I., Keller, C. A., Cromar, K. R., Pawson, S., & Ensz, H. (2021). Augmenting the standard operating procedures of health and air quality stakeholders with NASA resources. *Geohealth, 5*. https://doi.org/10.1029/2021GH000451.

European Space Agency: TROPOMI Level 2 Nitrogen Dioxide (2021). https://sentinels.copernicus.eu/web/sentinel/data-products/-/asset_publisher/fp37fc19FN8F/content/sentinel-5-precursor-level-2-nitrogen-dioxide. https://doi.org/10.5270/S5P-s4ljg54.

FLUID interactive map page for GEOS Composition Forecasts, https://fluid.nccs.nasa.gov/cf_map/index, last accessed 2023/05/04.

FLUID visualization page for GEOS Composition Forecasts. Retrieved May 4, 2023, from https://fluid.nccs.nasa.gov/cf/.

Giordano, M. R., Malings, C., Pandis, S. N., Presto, A. A., McNeill, V. F., Westervelt, D. M., Beekmann, M., & Subramanian, R. (2021). From low-cost sensors to high-quality data: A summary of challenges and best practices for effectively calibrating low-cost particulate matter mass sensors. *Journal of Aerosol Science., 158*, 105833. https://doi.org/10.1016/j.jaerosci.2021.105833

Google Earth Engine Data Catalog for GEOS-CF rpl tavg1hr v1: Goddard Earth Observing System Composition Forecast. Retrieved May 4, 2023, from https://developers.google.com/earth-engine/datasets/catalog/NASA_GEOS-CF_v1_rpl_tavg1hr.

Katoto, P. D. M. C., Byamungu, L., Brand, A. S., Mokaya, J., Strijdom, H., Goswami, N., De Boever, P., Nawrot, T. S., & Nemery, B. (2019). Ambient air pollution and health in Sub-Saharan Africa: Current evidence, perspectives and a call to action. *Environmental Research, 173*, 174–188. https://doi.org/10.1016/j.envres.2019.03.029

Keller, C. A., Evans, M. J., Knowland, K. E., Hasenkopf, C. A., Modekurty, S., Lucchesi, R. A., Oda, T., Franca, B. B., Mandarino, F. C., Díaz Suárez, M. V., Ryan, R. G., Fakes, L. H., & Pawson, S. (2021b). Global impact of COVID-19 restrictions on the surface concentrations of nitrogen dioxide and ozone. *Atmospheric Chemistry and Physics, 21*, 3555–3592. https://doi.org/10.5194/acp-21-3555-2021

Keller, C. A., Knowland, K. E., Duncan, B. N., Liu, J., Anderson, D. C., Das, S., Lucchesi, R. A., Lundgren, E. W., Nicely, J. M., Nielsen, E., Ott, L. E., Saunders, E., Strode, S. A., Wales, P. A., Jacob, D. J., & Pawson, S. (2021). Description of the NASA GEOS Composition Forecast Modeling System GEOS-CF v1.0. *Journal of Advances in Modeling Earth Systems, 13*. https://doi.org/10.1029/2020MS002413.

Malings, C., Westervelt, D. M., Hauryliuk, A., Presto, A. A., Grieshop, A., Bittner, A., Beekmann, M., Subramanian, R. (2020). Application of low-cost fine particulate mass monitors to convert satellite aerosol optical depth to surface concentrations in North America and Africa. *Atmospheric Measurement Techniques, 13*, 3873–3892. https://doi.org/10.5194/amt-13-3873-2020.

Malings, C., Knowland, K. E., Keller, C. A., & Cohn, S. E. (2021). Sub-city scale hourly air quality forecasting by combining models, satellite observations, and ground measurements. *Earth Space Science, 8*. https://doi.org/10.1029/2021EA001743.

McFarlane, C., Isevulambire, P. K., Lumbuenamo, R. S., Ndinga, A. M. E., Dhammapala, R., Jin, X., McNeill, V. F., Malings, C., Subramanian, R., & Westervelt, D. M. (2021a). First Measurements of Ambient $PM_{2.5}$ in Kinshasa, Democratic Republic of Congo and Brazzaville, Republic of Congo Using Field-calibrated Low-cost Sensors. *Aerosol and Air Quality Research, 21*. https://doi.org/10.4209/aaqr.200619.

McFarlane, C., Raheja, G., Malings, C., Appoh, E. K. E., Hughes, A. F., & Westervelt, D. M. (2021b). Application of Gaussian mixture regression for the correction of low cost $PM_{2.5}$ monitoring data in Accra, Ghana. *ACS Earth and Space Chemistry*, 1c00217. https://doi.org/10.1021/acsearthspacechem.1c00217.

NASA Applied Sciences ARSET Health & Air Quality Trainings. Retrieved May 5, 2023, from http://appliedsciences.nasa.gov/what-we-do/capacity-building/arset/arset-health-air-quality-trainings.

NASA Health and Air Quality Applied Sciences Team. Retrieved May 4, 2023, from https://haqast.org/.

OpenAQ Homepage. Retrieved May 5, 2023, from https://openaq.org/.

Open-Source Science Initiative. Retrieved May 5, 2023, from https://science.nasa.gov/open-science-overview.

Raheja, G., Sabi, K., Sonla, H., Gbedjangni, E. K., McFarlane, C. M., Hodoli, C. G., & Westervelt, D. M. (2022). A network of field-calibrated low-cost sensor measurements of $PM_{2.5}$ in Lomé, Togo, Over One to Two Years. *ACS Earth and Space Chemistry, 6*, 1011–1021. https://doi.org/10.1021/acsearthspacechem.1c00391.

Snyder, E. G., Watkins, T. H., Solomon, P. A., Thoma, E. D., Williams, R. W., Hagler, G. S. W., Shelow, D., Hindin, D. A., Kilaru, V. J., & Preuss, P. W. (2013). The Changing Paradigm of Air Pollution Monitoring. *Environmental Science & Technology., 47*, 11369–11377. https://doi.org/10.1021/es4022602

Sonoma Technology: Near-Real-Time Satellite Estimates of Surface PM2.5 Concentrations in Africa. Retrieved May 4, 2023, from https://www.sonomatech.com/projects/2881, last accessed 2023/05/04.

Subramanian, R., Kagabo, A. S., Baharane, V., Guhirwa, S., Sindayigaya, C., Malings, C., Williams, N. J., Kalisa, E., Li, H., Adams, P., Robinson, A. L., Langley DeWitt, H., Gasore, J., & Jaramillo, P. (2020). Air pollution in Kigali, Rwanda: spatial and temporal variability, source contributions, and the impact of car-free Sundays. *Clean Air Journal, 30*. https://doi.org/10.17159/caj/2020/30/2.8023.

World resources institute, resource watch webpage for air quality: Fine particulate matter ($PM_{2.5}$) station forecasts. Retrieved May 4, 2023, from https://resourcewatch.org/data/explore/cit004rw0-City-AQ-Forecasts.

Assessment of Urban Heat Island and Urban Pollution Island Synergy During the Dry Season in Greater Accra, Ghana

Cosmos S. Wemegah, Victoria Owusu Tawiah, Daniel M. Westervelt and Emmanuel K. Nyantakyi

Abstract

Accra is the capital and most populated region in Ghana but the smallest in the land coverage area. People living within the region could be highly exposed to the worse impacts of poor environmental conditions. Urban heat island (UHI) and pollution island (UPI) are two predicaments of the urban environment that are exacerbated by rapid urbanization. It is crucial to understand the synergy between UHI and UPI. In this paper, the interaction between UHI and UPI was studied by analyzing the diurnal variation of air temperature and fine particulate matter ($PM_{2.5}$) from purple-air sensors at six different sites in the capital. The observation revealed the existence of UHI and UPI with varying intensities. The UHI was nocturnal while UPI was more pronounced during daytime than night-time. The synergy analysis revealed some diurnal disparities which showed that UHI enhances UPI during the day while a coupling interaction at night-time resulted in strengthened nocturnal UHI in Accra. However, there were discrepancies in the interaction and the strength of the relationship at different sites. In summary, the synergy between UHI and UPI varies with location and may depend on the time of day, prevailing weather conditions, and other anthropogenic emissions. The findings also demonstrate the importance of addressing indiscriminate land use and necessitate the enforcement of existing urban planning laws within the country to mitigate the effect of air pollution amidst climate change and frequent extreme events which may exacerbate risk impact.

Keywords

Urban pollution Island · Urban heat island · Ghana air quality

1 Introduction

The intensive alteration of land use and air pollution emissions are two significant human activities in urban settings (McDonnell & MacGregor-Fors, 2016). The modification of the natural environment to impermeable surfaces results in greater energy storage in the surfaces which are more pronounced in urban areas. The urban areas, therefore, produce more excess warming than the suburban/rural areas resulting in a phenomenon called urban heat island (UHI) (Wemegah et al., 2020). Again, daily human activities such as the burning of fossil fuels in factories, automobiles, and other means of transport contribute significantly to the release of air pollutants into the atmosphere. This leads to the formation of an urban pollution island (UPI) where a nimiety of emissions occurs in the urban settings than in the surrounding suburban/rural (Crutzen, 2004). The UHI and UPI, therefore, increase the exposure of urban settlers to heat stress and air pollution with adverse health-related hazards (Burkart et al., 2013).

Several studies have found coupling interactions between UHI and UPI. This includes the enhancement of turbulent mixing by warm temperatures associated with UHI which in turn facilitates pollutant dispersion (Li et al., 2018; Sarrat et al., 2006). The increase in temperature could also boost pollutant concentration through chemical formation (Sarrat et al., 2006). Greater radiation forcing may be

C. S. Wemegah · E. K. Nyantakyi
University of Energy and Natural Resources, Earth Observation Research and Innovation Centre, Fiapre-Sunyani, Ghana
e-mail: cosmos.wemegah@uenr.edu.gh

V. Owusu Tawiah
Department of Meteorology and Climate Science, Kwame Nkrumah University of Science and Technology, Kumasi, Ghana

D. M. Westervelt
Columbia University, Lamont-Doherty Earth Observatory, 61 Route 9W, Palisades, NY 10964, USA

K. W. Fomba et al. (eds.), *Advances in Air Quality Research in Africa*,
Advances in Science, Technology & Innovation, https://doi.org/10.1007/978-3-031-53525-3_14

produced by high concentrations of pollutants in the atmosphere which could reduce incoming solar radiation and reduce temperatures (Wu et al., 2017). The intimate interaction between UHI and UPI could worsen the impacts of the phenomena on human life and activities. These effects vary at regional and local scales. Even though previous researchers have focused on UHI (Wemegah et al., 2020) and air pollutants (Owusu Tawiah et al., 2021; Sewor & Obeng, 2021) in Ghana, none explored UHI and UPI interactions. It is, therefore, crucial to comprehensively examine and understand the synergy between UHI and UPI in Accra, the capital of Ghana, hence the aim of this study.

2 Method

This study collected fine particulate matter ($PM_{2.5}$) and air temperature data from the Purple-Air low-cost sensors at six (6) sites in Accra, the capital of Ghana for the dry season of 2020. These datasets are made available online through the Ghana Urban Air Quality (GHAir) project. Details can be found in Sewor and Obeng (2021). The sites include Accra city hotel (ACH), Atomic Junction (AJ), Dzorwulu Junction (DJ), Roman Ridge (RR), Tetteh Quarshie interchange (TQI), all considered suburban sites and Ring Road Central (RRC) which was classified as an urban site due to its land-use cover. A correction factor developed by McFarlane et al. (2021) based on a colocation of Purple-Air at the US embassy was applied to correct the $PM_{2.5}$ concentrations. The differences between urban and suburban air temperature and the corrected $PM_{2.5}$ were used to determine the UHI and UPI intensities respectively with the equations below:

$$\text{UHII} = T_{\text{urban}} - T_{\text{suburban}} \tag{1}$$

$$\text{UPII} = \text{PM}_{\text{urban}} - \text{PM}_{\text{suburban}} \tag{2}$$

UHII and UPII are the intensities of UHI and UPI respectively. T_{urban} and $T_{suburban}$ are air temperatures for urban and suburban areas respectively while PM_{urban} is $PM_{2.5}$ for the urban site and $PM_{suburban}$ is $PM_{2.5}$ for suburban areas. The relationship between UHII and UPII was analyzed using Spearman's rank correlation.

3 Results and Discussion

The variation in diurnal air temperature $PM_{2.5}$, UHII and UPII for each site are presented in Fig. 1. In conformity with the initial prevision, RRC was indeed warmer than the other sites classified as suburban as observed in Fig. 1a with a distinct UHI feature more pronounced at night (see Fig. 1b). This is consistent with the findings of Wemegah et al. (2020) who found a noticeable nocturnal UHI with faster warming due to urbanization in the capital. Similar variations were observed in $PM_{2.5}$ with RRC recording relatively higher concentrations as shown in Fig. 1c and corroborate with Li et al. (2018) who reported higher PM concentrations in urban backgrounds than the suburban areas. However, Fig. 1d. revealed a dominant daytime UPI with the intensity peaking between 0600 and 0700 h.

The UHII correlated positively with UPII during the day across most sites except AJ with a strong negative correlation existing between nocturnal UHII and UPII except for ACH (details in Table 1.). The positive daytime correlation

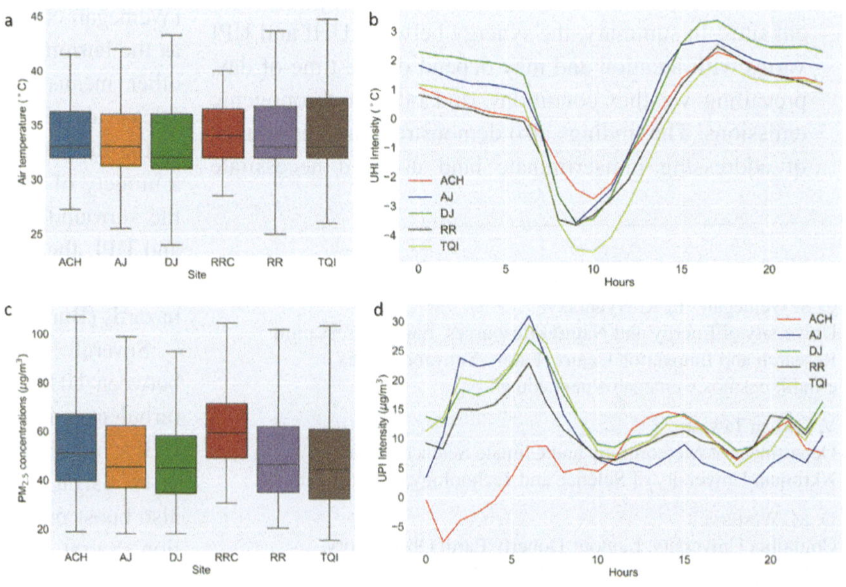

Fig. 1 Diurnal variation in air temperature **a** UHI intensities **b** $PM_{2.5}$ concentrations **c** and UPI intensities **d** for each site. UHI and UPI are determined with reference to RCC as the urban site

Table 1 Correlation between UHII and UPII for each site using RRC as the urban referenced site

Site	Day-time correlation	Night-time correlation
ACH	0.64	0.69
AJ	−0.31	−0.87
DJ	0.22	−0.92
RR	0.20	−0.75
TQI	0.24	−0.91

can be attributed to UHI warming enhancing chemical reactions and accumulation of pollutants and subsequently increasing the concentration levels (Sarrat et al., 2006). The impact of anthropogenic emissions during the day could however account for the weak positive correlation (Li et al., 2018). The negative correlation between UHII and UPII was consistent with previous studies by Cao et al. (2016) and Li et al. (2018). This could suggest that $PM_{2.5}$ enhances nocturnal UHI by trapping and re-emitting back longwave radiations to the urban surface. The dry season is characterized by prevailing hazy/dusty conditions transported from the Sahara Desert which could enhance nocturnal UHI as reported by Cao et al. (2016). A coupling effect where night-time UHI weakens UPII by facilitating aerosol particle dispersion through turbulent mixing has also been reported by Li et al. (2018).

4 Conclusion

This study investigated the synergy between UHI and UPI. The results revealed the existence of nocturnal UHI and a more pronounced daytime UPI within Accra. It was observed that daytime UHI strengthened UPI while nocturnal UHI weakened UPI which could be attributed to strong turbulent mixing. The results indicate that rapid urbanization coupled with extreme weather events such as heatwaves could worsen the risk impacts of UHI and poor air quality in Accra. It should, however, be noted that this study only considered atmospheric UHI within the dry season with distinct prevailing weather conditions, and therefore the synergy may differ with seasons, location and geographical features.

Acknowledgements The authors would like to acknowledge community members of the Ghana Urban Air Quality (GHAir) Project who hosted and made available Purple-Air low-cost sensor data at Accra City Hotel, Atomic Junction, Dzorwulu Junction, Roman Ridge, Ring Road Central, and Tetteh Quarshie Interchange.

References

Burkart, K., Canário, P., Breitner, S., Schneider, A., Scherber, K., Andrade, H., Alcoforado, M. J., & Endlicher, W. (2013). Interactive short-term effects of equivalent temperature and air pollution on human mortality in Berlin and Lisbon. *Environmental Pollution, 183*, 54–63.

Cao, C., Lee, X., Liu, S., Schultz, N., Xiao, W., Zhang, M., & Zhao, L. (2016). Urban heat islands in China enhanced by haze pollution. *Nature Communications, 7*(1), 12509.

Crutzen, P. J. (2004). New directions: The growing urban heat and pollution "island" effect-impact on chemistry and climate. *Atmospheric Environment, 38*(21), 3539–3540.

Li, H., Meier, F., Lee, X., Chakraborty, T., Liu, J., Schaap, M., & Sodoudi, S. (2018). Interaction between urban heat island and urban pollution island during summer in Berlin. *Science of the Total Environment, 636*, 818–828.

McDonnell, M. J., & MacGregor-Fors, I. (2016). The ecological future of cities. *Science, 6288*, 936–938.

McFarlane, C., Raheja, G., Malings, C., Appoh, E. K., Hughes, A. F., & Westervelt, D. M. (2021). Application of gaussian mixture regression for the correction of low cost $PM_{2.5}$ monitoring data in Accra Ghana. *ACS Earth and Space Chemistry, 5*(9), 2268–2279.

Owusu Tawiah, V., Westervelt, D., Annor, T., Hughes, A. F., & Lamptey, B. L. (2021). *Seasonal variation analysis of air pollutants in Accra-Ghana*. AGU.

Sarrat, C., Lemonsu, A., Masson, V., & Guedaliac, D. (2006). Impact of urban heat island on regional atmospheric pollution. *Atmospheric Environment, 40*(10), 1743–1758.

Sewor, C., & Obeng, A. A. (2021). The Ghana urban air quality project (ghair): Bridging air pollution data gaps in Ghana. *Clean Air Journal, 31*(1), 1–2.

Wemegah, C. S., Yamba, E. I., Aryee, J. N., Sam, F., & Amekudzi, L. K. (2020). Assessment of urban heat island warming in the greater accra region. *Scientific African, 8*, e00426.

Wu, H., Wang, T., Riemer, N., Chen, P., Li, M., & Li, S. (2017). Urban heat island impacted by fine particles in Nanjing China. *Scientific Reports, 7*(1), 1–11.

Recent Advances on Air Quality Monitoring and Modelling in Nigeria: Challenges and Future Prospects

Clement Kehinde Ajani, Oluwabamise Lekan Faboya, Khanneh Wadinga Fomba and Olubunmi Omotola Faboya

Abstract

Ambient air pollution is a global health threat that causes severe mortality, with impacts more intense in developing countries with large populations such as Nigeria. As such, air quality (AQ) monitoring and modelling are valuable techniques employed for investigating air pollution and its attendant problems. Hence, recent advances, main research gaps, challenges, and prospects for the use of these technologies in Nigeria are discussed based on published literature and other databases. The results revealed that current air assessment technologies in Nigeria are grossly limited. Thus, advanced approaches using satellite observations with high-resolution emission inventories, and state-of-the-art models for cross-scale and cross-regional air pollution impact assessments are required. To achieve these, advanced AQ surveillance and further research through cooperation with advanced nations, improved government policies, better regulations, and regional cooperation are strongly suggested.

Keywords

Air pollution · Air quality monitoring and modelling · Public health · Nigeria

1 Introduction

Globally, there have been increasing air pollution-related problems. Specifically, the status of atmospheric air pollution in Nigeria as a developing and most populous African country is quite alarming, with about 94% of Nigeria's population exposed to air pollution (World Bank, 2021). The associated health risks include; cardiovascular and respiratory diseases such as stroke, myocardial infarction, chronic pulmonary disease, and cancer. In particular, particulate matters ($PM_{2.5}$ and PM_{10}) have been linked with diabetes and a high risk of mortality in Covid-19 patients (Alvarez et al., 2020).

Despite these health-related risks, research on air quality (AQ) monitoring has been limited in Nigeria. This is partly due to inadequate AQ monitoring and modelling technologies in Nigeria (Alvarez et al., 2020). On the other hand, a relevant literature review on independent AQ studies in Nigeria carried out by Offor et al. (2016), which spanned from 1985 to 2015, was only focused on particulate matter and elemental composition of aerosols. Consequently, the current study provides updated knowledge on the recent advances in AQ monitoring and modelling technologies in Nigeria, reviews current trends on AQ technologies employed, highlights challenges and future research prospects, and proffers cooperation strategies.

2 Materials and Methods

The current review was conducted using research and review papers obtained from different bibliographical databases such as "Web of Science" (WOS), "Scopus" (Sc),

C. K. Ajani (✉)
African Research Centre on Air Quality and Climate, University Mohammed VI Polytechnic, 43150 Benguerir, Morocco
e-mail: clemajan@gmail.com

O. L. Faboya
Department of Chemistry, Federal University Oye-Ekiti, Ekiti State, Oye-Ekiti, Nigeria

O. L. Faboya · K. W. Fomba
Atmospheric Chemistry Department (ACD), Leibniz Institute for Tropospheric Research (TROPOS), Permoserstraße 15, 04318 Leipzig, Germany

O. O. Faboya
Department of Computer Science, Bamidele Olumilua University of Education, Science, and Technology, Ikere-Ekiti, Nigeria

K. W. Fomba et al. (eds.), *Advances in Air Quality Research in Africa*,
Advances in Science, Technology & Innovation, https://doi.org/10.1007/978-3-031-53525-3_15

Table 1 Summary of AQ monitoring and modelling techniques in reviewed studies

Monitoring techniques/models	Location/findings	References
Low-cost sensors and IoT/statistical model	Lagos/pollutants higher than WHO guidelines	Abulude et al. (2021)
SPARTAN/ANN	Ilorin/model was well correlated with measured data ($p = 0.0009$, $R^2 = 0.59$–0.95)	Onyeuwaoma et al. (2021)
Altair 5× multi-gas detector/Gaussian model	Kano/model suitable for forecasting CO & NO_2 emissions	Tyovenda et al., (2021)
Gas monitor (sensors), GPS, ArcGIS/ mass balance model	Port Harcourt/concentrations above FMEnv and NAAQS limits	Ogaji et al. (2022)
ISCST3 dispersion models	Kaduna, Port Harcourt, Warri/NO_x and SO_2 exceeded the set limit from modelling	Oladimeji et al. (2015)
Mobile AQM device (Gasman air and haze-dust particulate monitors, GPS map equipment)/ statistical models	Orlu/AQ index withing 151–225 implying unhealthy atmosphere	Chizoruo et al. (2017)
Low-cost monitoring sensors/statistical model	Ile-Ife/Daily peaks of $PM_{2.5}$, PM_{10}, and CO_2, were mostly in the morning while the peak period of CO, NO_2, and NO was in the evening	Owoade et al. (2021)
Absorption trains/statistical model	Ibadan/SO_2 exceeded the WHO limit	Ipeaiyeda and Adegboyega (2017)

National ambient air quality standards: NAAQS; federal ministry for the environment: FMEnv; World Health Organization: WHO; industrial source complex short term: ISCST3; surface particulate matter network (SPARTAN), artificial neural network: ANN; air quality monitoring: AQM; gas-sensitive electrochemical: GSE; gas-sensitive semi-conductor: GSS

etc., and other sources including the WHO and World Bank. Moreover, the results were limited to 2015–2022, and the literature was centered on both outdoor and indoor air pollution. We searched using the following keywords "air pollution", "air quality monitoring", "air quality modelling", "government policies", and pollutants such as "particulate matter (PM)", "carbon monoxide (CO)", "nitrogen dioxide (NO_2)" etc., while limiting the search to Nigeria. In addition, only articles with measured or modelled air quality alongside the technologies used were included in the review.

3 Advances in Air Quality Monitoring and Modelling in Nigeria

A summary of the recent research on air quality monitoring (AQM) in Nigeria is presented in Table 1. From the table, it can be inferred that for the most part, relatively cheap and simple AQM devices were employed. These include low-cost sensors (Abulude et al., 2021; Ogaji et al., 2022; Owoade et al., 2021), portable mobile devices (Chizoruo et al., 2017), and absorption trains (Ipeaiyeda & Adegboyega, 2017). On the other hand, the AQ models were mostly limited to using basic statistical analysis, except for a few studies which used ANN (Onyeuwaoma et al., 2021), Gaussian model (Tyovenda et al., 2021), and ISCST3 dispersion model (Oladimeji et al., 2015), as displayed in Table 1.

4 Challenges and Future Directions

4.1 Air Quality Monitoring

From the current review, it can be inferred that Nigeria is lagging in monitoring AQ, thus indicating a sparsity of data on atmospheric compositions. Hence, research focus on the routine evaluation of aerosol, chemical, microphysical, and optical properties, of $PM_{2.5}$, PM_{10}, etc., should be intensified. Also, spatiotemporal surveillance of important parameters (e.g., aerosol composition) is needed (Offor et al., 2016). Currently, research on AQ monitoring in Nigeria has been limited to independent measurements mostly at a single scale, thus, AQ monitoring systems with multi-dimension, multi-parameter and multiscale measurements capabilities could be employed for future research. Measurements using remote sensing play an important role in AQ monitoring (Chizoruo et al., 2017), therefore, enhancing the spatial resolution of current remote sensing systems and improving their synergy with relevant modelling tasks is pertinent (Baklanov & Zhang, 2020). Moreover, mobile platforms are viable means of operating surface measuring instruments and can be explored (Apte et al., 2017). Hitherto, low-cost sensors have been used for monitoring AQ (Abulude et al., 2021), these sensors have high spatiotemporal variabilities and their quality is always not guaranteed. Hence, AQ systems with sensors that are accurate but economical should be employed. In addition, there are emerging low-cost sensing devices that could be

Table 2 Comparison between the present review and other related reviews

Review objective	Review findings	References
Review of air quality policies, monitoring, and health implications in West African	In most countries, air quality monitoring, legislation, and public health surveillance data are insufficient/unavailable	Alvarez et al. (2020)
Review of air quality pre-treatment, analytical and sampling methods in Nigeria (1985 – 2015)	PM loads are higher in dry season than rainy season and in urban areas than in rural areas	Offor et al. (2016)
Review of urban air quality modeling	Models are required to understand the pollutants' dispersion dynamics including their interaction with urban features	Kadaverugu et al. (2019)
Review of research advances on air pollution in 46 countries in Sub-Saharan Africa	Limited data on air pollutant exposures to non-communicable respiratory diseases in African	Glenn et al. (2022)
Review of air pollution and its effects on human capital, economy, and health across Africa (Ethiopia, Ghana, and Rwanda)	Air pollution effects in 2019:1·1 million deaths; economic loss of $3·02, billion, $1·63 billion, and $349 million in Ethiopia, Ghana, and Rwanda, respectively	Fisher et al. (2021)

integrated with existing compact ones for high-resolution detection of pollutants (Jaffe et al., 2014).

4.2 Air Quality Modelling

Apart from policies and regulations, researchers and relevant authorities need AQ mathematical models to devise scientific management plans for reducing associated impacts. As such, both simple and complex AQ models have been developed across different scales (Baklanov & Zhang, 2020). However, research on advanced AQ models such as chemical transport models (CTMs), computational fluid dynamics (CFD) models, mesoscale models, and their hybrids (Baklanov & Zhang, 2020; Kadaverugu et al., 2019), in Nigeria are lacking. Hence, a spectrum of these high-resolution models (CTMs, CFD, and Reynolds-averaged Navier–Stokes (RANS) are important for modelling AQ across varying scales (Kukkonen et al., 2020) and could be applied for relevant research. Similarly, machine learning (ML) approaches can be employed to improve measured data for accurate personal exposure and fine-scale AQ predictions (Shaddick et al., 2021). Additionally, ensemble simulation provides more reliable AQ prediction and can be combined with other approaches for multi-scale predictions (Galmarini et al., 2018). Current models from the review have no links with the associated health risk models, hence, such should be investigated. Lastly, models for evaluating and detecting the nature and amount of errors in measured data have been applied in recent research and can also be adopted (Casazza et al., 2019).

5 Comparative Study with Previous Reviews

As summarized in Table 2, previous relevant reviews were focused on AQ monitoring, policies, and its attendant issues across Africa (Alvarez et al., 2020; Fisher et al., 2021; Glenn et al., 2022), some on evaluating AQ models in urban settings (Kadaverugu et al., 2019), and others on AQ sampling and analytical methods in Nigeria (Offor et al., 2016). Unlike previous reviews, the current work focused on evaluating advances in AQ monitoring and modeling technologies in Nigeria from 2015 to 2022, hence, its novelty.

6 Conclusion

A review of recent research developments on air pollution monitoring and modelling in Nigeria is presented in the current study. In particular, the current status of air quality monitoring technologies and relevant air pollution models employed were assessed. Furthermore, emerging challenges and prospects for future research were highlighted. Results revealed that current air quality monitoring equipment and models are limited in evaluating air pollution status and attendant health risks in Nigeria. Besides implementing relevant policies and regulations, the use of advanced monitoring facilities and state-of-the-art models such as CFD models, CTMs, ML models, ensemble simulations, etc., are recommended. This will enhance more reliable prediction of air quality and its associated impacts on public health, ecosystem, and climate.

References

Abulude, F. O., Damodharan, U., Acha, S., Adamu, A., & Arifalo, K. M. (2021). Preliminary assessment of air pollution quality levels of Lagos Nigeria. *Aerospace Science and Engineering, 5*(3), 275–284.

Alvarez, C. M., Hourcade, R., & Lefebvre, B. (2020). A Scoping review on air quality monitoring policy and health in west African cities. *International Journal of Environmental Research and Public Health, 17*(9151), 1–26.

Apte, J. S., Messier, K. P., Gani, S., Brauer, M., Kirchstetter, T. W., Lunden, M. M., Marshall, J. D., Portier, C. J., Vermeulen, R. C. H., & Hamburg, S. P. (2017). High-Resolution air pollution mapping with google street view cars: Exploiting big data. *Environmental Science & Technology, 51*(12), 6999–7008.

Baklanov, A., & Zhang, Y. (2020). Advances in air quality modeling and forecasting Advances in air quality modeling and forecasting. *Global Transitions, 2*, 261–270.

Casazza, M., Lega, M., Jannelli, E., Minutillo, M., Jaffe, D., Severino, V., & Ulgiati, S. (2019). 3D monitoring and modelling of air quality for sustainable urban port planning: Review and perspectives. *Journal of Cleaner Production, 231*, 1342–1352.

Chizoruo, I. F., Alexander, O., Pascal Chu, N., & Jude Ikech, A. (2017). Ambient air quality assessment of Orlu, Southeastern Nigeria. *Journal of Applied Sciences, 17*(9), 441–457.

Fisher, S., Bellinger, D. C., Cropper, M. L., Kumar, P., Binagwaho, A., Koudenoukpo, J. B., Park, Y., Taghian, G., Landrigan, P. J.:Articles Air pollution and development in Africa : impacts on health , the economy , and human capital. Fisher, S., Bellinger, D. C., Cropper, M. L., Kumar, P., Binagwaho, A., Koudenoukpo, J. B., Park, Y., Taghian, G., & Landrigan, P. J. (2021). Articles Air pollution and development in Africa: impacts on health, the economy, and human capital. *The Lancet Planetary Health, 5*(10), e681–e688.

Galmarini, S., Kioutsioukis, I., Solazzo, … Unal, A. (2018). Two-scale multi-model ensemble: Is a hybrid ensemble of opportunity telling us more? *Atmospheric Chemistry and Physics, 18*, 8727–8744.

Glenn, B. E., Espira, L. M., Larson, M. C., & Larson, P. S. (2022). Ambient air pollution and non-communicable respiratory illness in sub-Saharan Africa : A systematic review of the literature. *Environmental Health, 21*(40), 1–14.

Ipeaiyeda, A. R., & Adegboyega, D. A. (2017). Assessment of air pollutant concentrations near major roads in residential, commercial and industrial areas in Ibadan City Nigeria. *Journal of Health and Pollution, 7*(13), 11–21.

Jaffe, D. A., Hof, G., Malashanka, S., Putz, J., Thayer, J., Fry, J. L., Ayres, B., & Pierce, J. R. (2014). Diesel particulate matter emission factors and air quality implications from in-service rail in Washington State USA. *Atmospheric Pollution Research, 5*, 344–351.

Kadaverugu, R., Sharma, A., Matli, C., & Biniwale, R. (2019). High Resolution Urban air quality modeling by coupling CFD and mesoscale models: A review Asia-Pacific. *Journal of the Atmospheric Sciences, 55*, 539–556.

Kukkonen, J., Savolahti, M., Palamarchuk, Y., Lanki, T., Nurmi, V., & Paunu, V. (2020). (2020) Modelling of the public health costs of fine particulate matter and results for Finland in 2015. *Atmospheric Chemistry and Physics, 20*, 9371–9391.

Offor, I. F., Adie, G. U., & Ana, G. R. (2016). Review of particulate matter and elemental composition of aerosols at selected locations in Nigeria from 1985–2015. *Journal of Health and Pollution, 6*, 1–18.

Ogaji, F. M., Obafemi, A., Numbere, A. O., & Ogaji, D. S. (2022). Geospatial analysis and modeling of indoor air quality in some residential areas in the Niger. *Low Carbon Economy, 13*, 1–24.

Oladimeji, T., Sonibare, J., Odunfa, M., & Ayeni, A. (2015). Modeling of criteria air pollutant emissions from selected Nigeria petroleum refineries. *Journal of Power and Energy, 3*, 31–45.

Onyeuwaoma, N., Okoh, D., & Okere, B. (2021). A neural network-based method for modeling $PM_{2.5}$ measurements obtained from the surface particulate matter network. *Environmental Monitoring and Assessment, 193*, 1–15.

Owoade, O. K., Abiodun, P. O., Omokungbe, O. R., Fawole, O. G., Olise, F. S., Popoola, O. O. M., Jones, R. L., & Hopke, P. K. (2021). Spatial-temporal variation and local source identification of air pollutants in a semi-urban settlement in Nigeria using low-cost sensors. *Aerosol and Air Quality Research, 21*(10), 1–18.

Shaddick, G., Salter, J. M., Peuch, V.-H., Ruggeri, G., Thomas, M. L., Mudu, P., Tarasova, O., Baklanov, A., & Gumy, S. (2021). Global air quality: An inter-disciplinary approach to exposure assessment for burden of disease analyses. *Atmosphere, 12*(48), 1–13.

Tyovenda, A. A., Ayua, T. J., & Sombo, T. (2021). Modeling of gaseous pollutants (CO and NO_2) emission from an industrial stack in Kano city, northwestern Nigeria. *Atmospheric Environment, 253*, 118356.

World Bank. (2021). A plea for action against pollution in Nigeria. Retrieved July 12, 2021, from http://www.worldbank.org/en/news.

Relationships Between Meteorological Parameters and $PM_{2.5}$ in Accra

Victoria Owusu-Tawiah, Daniel M. Westervelt and Thompson Annor

Abstract

Studies around the world have associated fine particulate matter ($PM_{2.5}$) with human health. The concentrations of $PM_{2.5}$ is influenced by multiple meteorological conditions as well as socioeconomic and geographical factors. Using $PM_{2.5}$ obtained from Purple Air sensors and Beta Attenuation Monitor (BAM), this study assessed the variation in $PM_{2.5}$ in different time steps and investigated the relationship with meteorological variables. The results showed a monthly variation in $PM_{2.5}$ with higher concentration and poor air quality observed during February which could be attributed to the prevailing dusty condition from the Sahara Desert experienced in Ghana during the dry season. September recorded lower concentrations with good to moderate levels of air quality concerns since this period falls within the wet season. A positive correlation existed between $PM_{2.5}$ and temperature while the correlation was negative for relative humidity and wind speed. This suggests that increasing temperature enhances the concentrations of $PM_{2.5}$ while higher relative humidity weakens the concentrations. The relationship between $PM_{2.5}$ and wind speed also indicated that calm winds enhance the concentrations while fast winds disperse the pollutants.

Keywords

Particulate matter 2.5 · Air quality index · Accra Ghana

V. Owusu-Tawiah (✉) · T. Annor
Department of Physics, Kwame Nkrumah University of Science and Technology, Kumasi, Ghana
e-mail: vikowusu@gmail.com

D. M. Westervelt
Columbia University, Lamont-Doherty Earth Observatory, 61 Route 9W, Palisades, NY 10964, USA

1 Introduction

In recent times, serious efforts have been geared towards combating particulate matter (PM) pollution. Particulate matter is associated with numerous cardiovascular and respiratory complications (Hime et al., 2018). Particulate matter with an aerodynamic size of 2.5 µm or less ($PM_{2.5}$) has continued to pose a significant threat to human health around the world, especially in developing countries. The level of air pollution in a given area depends on various factors, including the type of fuel being consumed, and geographical and topographical features. Meteorology and atmospheric chemistry can also significantly influence the quality of air in cities (Manju et al., 2018; Zhang et al., 2015). Parameters such as wind speed and direction, temperature, or relative humidity can change the process of pollutant dispersion, deposition mechanisms, and the formation of atmospheric particles and thus influence the concentration of air pollutants in a region (Zhang et al., 2015).

Rapid urbanization and the existence of UHI has been observed in Accra, the capital city of Ghana (Wemegah et al., 2020), and are among the key driving force of poor air quality (Mayer, 1999) and are known to exacerbate the effects of air pollutants in urban areas (Wemegah et al., 2020). Again, the annual reported $PM_{2.5}$ mean concentration in Ghana in 2016 was 31.1 $\mu g/m^3$ which exceeded the recommended standard level of 10 $\mu g/m^3$ (World Health Organization, 2016). This gives an indication of the growing poor air quality in Ghana and its subsequent well-known consequences on human health (Odonkor & Mahami, 2020). A report from the Global Burden of Disease study in 2016 suggested that exposure to poor air quality has led to about 28,000 premature deaths in Ghana (World Health Organization, 2016). There is, therefore, the need to study and analyze the state of air quality in Ghana to improve accuracy and reliable air quality forecasts. Most work done in Ghana focused on ambient PM concentrations in the Greater Accra region, where the sources of air

K. W. Fomba et al. (eds.), *Advances in Air Quality Research in Africa*,
Advances in Science, Technology & Innovation, https://doi.org/10.1007/978-3-031-53525-3_16

pollutants have been identified (Arku et al., 2008). This work therefore aims to investigate the impact of meteorological parameters (temperature, relative humidity, wind speed, and direction) on particulate matter concentration in Accra over a 12-month observation period.

2 Method

The study utilizes real-time hourly concentrations of $PM_{2.5}$ from a Beta-Attenuation Monitor (BAM) situated at the US Embassy in Accra, along with low-cost purple air sensors (PAS) positioned at four (4) different locations within the Greater Accra Metropolitan Area. The data collection period spans from March 2020 to February 2021. The purple air PM concentrations were subjected to a correction factor as determined by McFarlane et al. (2021) and quality controlled. These were then averaged to form a point time series $PM_{2.5}$ data for Accra. The aggregated data was compared with the BAM data to assess the consistency between the two datasets.

The air quality index (AQI) was calculated from the hourly concentrations of PM data before aggregating to daily, monthly, and seasonal time steps. The aim of this was to explore how meteorology influences PM concentration and air quality at various time intervals. Meteorological data for the study area which includes temperature, relative humidity, wind speed, and direction, were obtained from the Ghana Meteorological Agency (GMet) through the Global Telecommunication System (GTS). The correlation between meteorological parameters and PM concentrations was analyzed through the application of Spearman-Rank correlation and the simple regression method.

3 Results and Discussions

The monthly means of pollutant concentrations and AQI were analyzed and presented in Fig. 1. The PAS and BAM $PM_{2.5}$ varied between 18.08–57.88 $\mu g\ m^{-3}$ and 16.54–110.93 $\mu g\ m^{-3}$ respectively. Even though monthly concentrations were generally higher in January for all PMs, the peaks were recorded in February. AQI indicated mostly good to moderate air quality levels were observed throughout with hikes in concerns levels (unhealthy to hazardous) experienced mostly in February. BAM however recorded unhealthy levels in January and December.

High $PM_{2.5}$ concentrations recorded from December to February can be attributed to the Harmattan season settling in during the same period. This season is generally characterized by the influx of dust from the Sahara Desert (Nicholson, 2013). This influx of dust coupled with high temperatures and low humidity causes hazy conditions across the entire country during this season thereby affecting the quality of air.

The meteorological variables also exhibited bimodal variation for relative humidity (RH) and unimodal patterns for temperature and wind speed (WS) as shown in Fig. 2.

The air temperatures ranged from 29.03 to 34.48 °C with a peak in January and recorded the lowest in August. RH varied throughout the year with a noticeable peak in September and recorded the lowest levels in February while a minor and a major peak in WS were observed in March and August respectively. Higher and lower temperatures observed in January and August respectively confirm the positive correlation of 0.55 between temperature and $PM_{2.5}$. The RH and WS however reversely correlated (−0.84 and −0.61 respectively) with $PM_{2.5}$ concentrations.

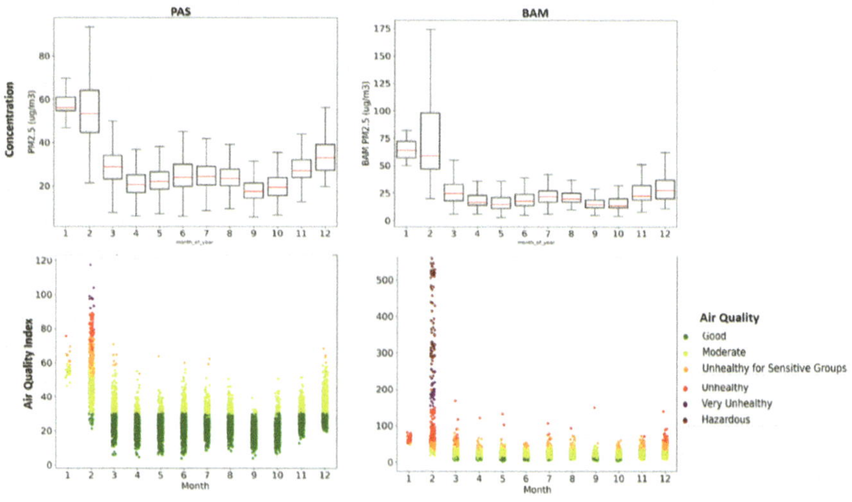

Fig. 1 Monthly variation in $PM_{2.5}$ concentrations and air quality index from PAS and BAM

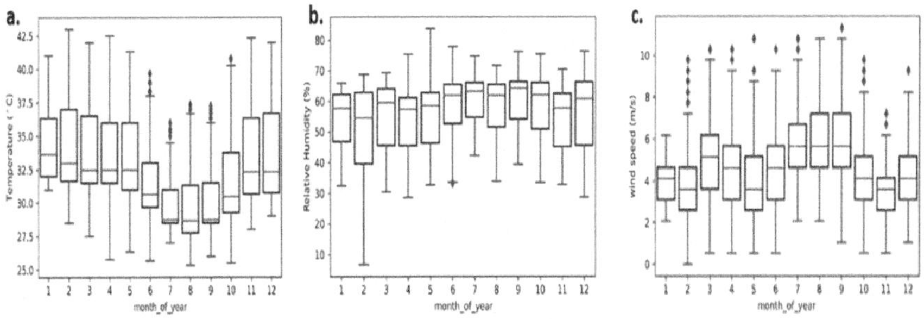

Fig. 2 Monthly variation in meteorological variables: **a** temperature; **b** RH; **c** wind speed

4 Conclusions

In this study, the connection between meteorological parameters (temperature, relative humidity, wind speed, and direction) and concentrations of fine particulate matter ($PM_{2.5}$) was investigated using data acquired from Purple Air sensors and a federal monitor in the Greater Accra Metropolitan Area. The findings showed that the city experienced poor air quality within the period of study. February recorded the highest concentrations with very poor air quality. This was attributed to the prevailing dusty conditions during the dry season. A positive correlation existed between $PM_{2.5}$ and temperature which indicated that increasing temperature supports poor air quality. It is however recommended that harmattan days be removed from the months with high concentrations in further studies to determine other sources contributing to the high $PM_{2.5}$ concentrations observed.

Acknowledgements The authors would like to acknowledge the Ghana Meteorological Agency, the US Embassy in Accra, and the Ghana AQ project for providing access to the various datasets used in this study.

References

Arku, R. E., Vallarino, J., Dionisio, K. L., Willis, R., Choi, H., Wilson, J. G., Hemphill, C., Agyei-Mensah, S., Spengler, J. D., & Ezzati, M. (2008). Characterizing air pollution in two low-income neighborhoods in Accra Ghana. *Science of the Total Environment, 402*(2–3), 217–231.

Hime, N. J., Marks, G. B., & Cowie, C. T. (2018). A comparison of the health effects of ambient particulate matter air pollution from five emission sources. *International Journal of Environmental Research and Public Health, 15*(6), 1206.

Manju, A., Kalaiselvi, K., Dhananjayan, V., Palanivel, M., Banupriya, G. S., Vidhya, M. H., Panjakumar, K., & Ravichandran, B. (2018). Spatio-seasonal variation in ambient air pollutants and influence of meteorological factors in Coimbatore, Southern India. *Air Quality, Atmosphere & Health, 11*(10), 1179–1189.

Mayer, H. (1999). Air pollution in cities. *Atmospheric Environment, 33*(24–25), 4029–4037.

McFarlane, C., Raheja, G., Malings, C., Appoh, E.K., Hughes, A.F. and Westervelt, D. M. (2021). Application of Gaussian mixture regression for the correction of low cost $PM_{2.5}$ monitoring data in Accra, Ghana. *ACS Earth and Space Chemistry, 5*(9), 2268–2279.

Nicholson, S. E. (2013). The West African Sahel: A review of recent studies on the rainfall regime and its interannual variability. *International Scholarly Research Notices*.

Odonkor, S. T., & Mahami, T. (2020). Knowledge, attitudes, and perceptions of air pollution in Accra, Ghana: A critical survey. *Journal of Environmental and Public Health*.

Wemegah, C. S., Yamba, E. I., Aryee, J. N., Sam, F., & Amekudzi, L. K. (2020). Assessment of urban heat island warming in the greater accra region. *Scientific African, 8*, e00426.

World Health Organization. (2016). *Ambient air pollution: A global assessment of exposure and burden of disease*. World Health Organization.

Zhang, H., Wang, Y., Hu, J., Ying, Q., & Hu, X. M. (2015). Relationships between meteorological parameters and criteria air pollutants in three megacities in China. *Environmental Research, 140*, 242–254.

Assessment of $PM_{2.5}$ in Blantyre City, Malawi

Fabiano Gibson Daud Thulu, Chikumbusko Chiziwa Kaonga, Ishmael Bobby Mphangwe Kosamu and Mathews Nyasulu

Abstract

Measuring and evaluating environmental levels of air pollution is very crucial for determining detrimental effects of air pollution to people working in sources of air pollution as well as the general public. In this study, $PM_{2.5}$ levels were measured in Malawi (Blantyre City). 36 short-term ground measurements were performed using an active mobile multi-gas portable monitor known as Dylos DC1100 PRO Laser Particle Counter, in 18 different locations for a period of 60 days in 2020. 3 monitoring points were selected from school campuses, hospitals, industrial areas, markets, residential areas and Central Business District (CBD) of Blantyre for this air quality investigation. $PM_{2.5}$ monitoring was conducted between 10:00–12:00 and 17:00–19:00. The result show that the values of $PM_{2.5}$ ranged from 2.51 to 27.69 μg/m^3, 1.67 to 24.81 μg/m^3, 5.74 to 26.69 μg/m^3, 2.72 to 42.57 μg/m^3, 1.31 μg/m^3 to 15.57 μg/m^3 and 3.54 μg/m^3 to 20.93 μg/m^3 for school campuses, hospitals, industrial areas, markets, residential areas and CBD respectively. The findings were compared with the public exposure limits of both local air quality Malawi Standards and on World Health Organisation standards particulate pollution of 25 μg/m^3 and 15 μg/m^3 for 24 h respectively. Significant higher exposures of $PM_{2.5}$ due to direct exposure to emissions from dirty solid fuel combustion and dusty storms were observed, indicating poor air quality. Long term and continuous monitoring of $PM_{2.5}$ is required over the city to establish concentration trends for decision making.

Keywords

$PM_{2.5}$ · Air quality · Pollution · Exposure · Malawi

1 Introduction

Industrialisation and environmental degradation are expected to continue in Africa, as such, the amount of air pollution will continue to increase steadily (Thulu et al., 2023). Air pollution is mostly a result of anthropogenic activities. The local sources of air pollution vary in magnitude, space, and time (Osipov et al., 2022). Air pollutions is the addition or introduction of various hazardous chemicals, toxic substances, particulate matter (PM), and biological organisms that harm or cause discomfort to humans or other living organisms, or damage the environment and atmosphere (Okello et al., 2022). By definition, $PM_{2.5}$ means particles that pass through a size-selective inlet with a 50% efficiency cut-off at 2.5 μm aerodynamic diameter. PM consists of a complex mixture of solid and liquid particles of inorganic and organic substances in the air (Xu et al., 2020). Larger particles are generally filtered in the nose and throat and generally cause no problems while finer particles (10 microns or less) can pass through and settle in the bronchi and lungs of the human body. These fine particles generally cause health problems and contribute to the risk of development of respiratory disease, cardiovascular disease, and lung cancer (Derafshi et al., 2017). In Malawi, little has been done to fully assess the $PM_{2.5}$ spatiotemporal variability on the credibility of using isolated ambient observations for regulatory purposes using ground monitored data.

From Malawi Specification MS737:2011, the limit value for the annual average concentration of $PM_{2.5}$ is 8 μg/m^3 and limit value for 24 h averages is 25 μg/m^3 (MBS, 2017). This information relates area-wide health statistics to area-wide air quality (Jary et al., 2015). The WHO recommended limits for the concentration of $PM_{2.5}$

F. G. D. Thulu (✉) · C. C. Kaonga · I. B. M. Kosamu
Physics Department, Malawi University of Business and Applied Sciences, Private Bag 303, Chichiri, Blantyre 3, Malawi
e-mail: fthulu@poly.ac.mw

M. Nyasulu
School of Applied Meteorology, Nanjing University of Information Science and Technology, Nanjing 210044, China

K. W. Fomba et al. (eds.), *Advances in Air Quality Research in Africa*,
Advances in Science, Technology & Innovation, https://doi.org/10.1007/978-3-031-53525-3_17

is ≤ 15 μg/m^3 averaged during the 24-h period outside the facility boundary, and it cannot be exceeded 35 times in a year (Organización Mundial de la Salud (OMS), 2021).

$PM_{2.5}$ is originated from many sources, both natural the anthropogenic derivatives (Cohen et al., 2004). Some of the natural sources include: forest fires, pollen scattering, sand storms blown up from the surface of the earth by strong winds, sea salt particles being blown up from the surface of the sea by winds, and ozone which has subsided from the stratosphere to the troposphere (Obstructive et al., 2018). Unlike natural sources of pollution, anthropogenic sources of pollution emit from mild to strong concentration of pollutants on a scale of weak-mild-strong (Thurlow et al., 2012).

Measuring, evaluating and determining the effects of environmental levels of $PM_{2.5}$ is more crucial to the occupation and the public. This is why, this topic is subject to the ongoing research in Malawi and it is vital to the optimization of air quality in Malawi. Many factors can cause change in air pollution. Measuring and evaluating the levels of air pollution in crowded places has become of utmost importance in Africa. There is limited research on $PM_{2.5}$ in Africa, with some countries like Malawi have no $PM_{2.5}$ ground researched values (Mapoma et al., 2013). In this study, concentration changes of $PM_{2.5}$ levels were conducted in 28 locations in school campuses, hospitals, industrial areas, markets, residential areas and Blantyre Central Business District (CBD) for a continuous 60-day period during morning and evening hours. This air quality monitoring exercise was aimed at identifying possible changes in $PM_{2.5}$. Results from this monitoring exercise may help to make necessary recommendation and decisions to ensure environmentally friendliness within the city of Blantyre.

2 Measurement of Concentration Changes of $PM_{2.5}$

This study aimed at monitoring and evaluating the concentration changes of $PM_{2.5}$ levels in school campuses, hospitals, industrial areas, markets, residential areas and CBD in Malawi. Sample collection was done using an active mobile multi-gas portable monitor (Dylos DC1100 PRO Laser Particle Counter) equipped with electrochemical sensors that measure continuous concentrations of various gaseous parameters of interest per 90 min session during each sampling period. At all the locations, the equipment was placed at 1.5 m from the ground. Each measurement at the same location lasted for 30 min and was repeated three times and the maximum $PM_{2.5}$ value was averaged and recorded. The measurements of $PM_{2.5}$ were made twice on the same day, one in the morning and another in the afternoon. The map of measurement locations selected in Blantyre city is presented in Fig. 1 below.

3 Measurement Results

From the ground monitored results of $PM_{2.5}$, it was found out that the mean values of $PM_{2.5}$ ranged from 2.51 to 27.69 μg/m^3, 1.67 to 24.81 μg/m^3, 5.74 to 26.69 μg/m^3, 2.72 to 42.57 μg/m^3, 1.31 μg/m^3 to 15.57 μg/m^3 and 3.54 μg/m^3 to 20.93 μg/m^3 for school campuses, hospitals, industrial areas, markets, residential areas and CBD respectively. Table 1 shows the recorded maximum $PM_{2.5}$ concentration value and computed mean values of the peck in the monitored areas for a 60-day period. The lowest $PM_{2.5}$

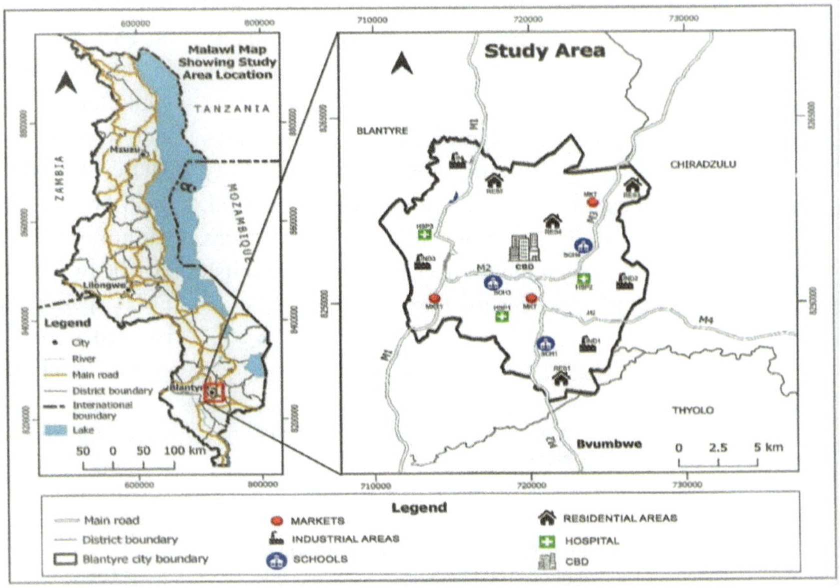

Fig. 1 Map of measurement locations in Blantyre City, Malawi

Table 1 Statistical results of dairy peck $PM_{2.5}$ concentrations in the monitored areas of Blantyre City, Malawi

Location	Coordinates (lat and long)	Measurements between 10:00 and 12 h (μg/m³)			Measurements between 17:00 and 19:00 h (μg/m³)		
		Max	Mean	Std	Max	Mean	Std
SCH 1	15⁰47′ 18 S 35⁰ 01′ 47 E	27.43	21.64	0.93	22.63	19.54	0.34
SCH 2	15⁰ 47′ 48 S 35⁰ 01′ 22 E	23.65	21.02	0.67	24.76	16.53	0.53
SCH 3	15⁰ 48′ 42 S 35⁰ 00′ 01 E	22.75	20.94	0.53	14.22	10.53	0.62
HSP 1	15⁰ 46′ 20 S 35⁰ 01′ 28 E	23.75	12.56	0.25	18.72	12.35	0.64
HSP 2	15⁰ 41′ 35 S 35⁰ 01′ 29 E	24.81	18.04	0.64	15.03	11.47	0.46
HSP 3	15⁰ 43′ 44 S 35⁰ 00′ 04 E	14.75	11.57	0.27	11.24	6.04	0.92
IND 1	15⁰ 47′ 32 S 35⁰ 01′ 04 E	19.96	14.64	0.53	25.74	20.99	0.52
IND 2	15⁰ 49′ 33 S 35⁰ 00′ 56 E	20.69	17.87	0.59	19.74	12.23	0.83
IND 3	15⁰ 46′ 42 S 35⁰ 01′ 52 E	25.27	10.57	0.24	14.75	10.83	0.64
MKT 1	15⁰ 49′ 05 S 35⁰ 01′ 22 E	42.83	15.42	0.54	22.43	11.64	0.36
MKT 2	15⁰ 48′ 12 S 34⁰ 59′ 32 E	31.53	13.53	0.42	23.56	14.24	0.55
MKT 3	15⁰ 43′ 23 S 35⁰ 59′ 47 E	28.64	12.93	0.35	24.24	17.53	0.37
RES 1	15⁰ 47′ 25 S 35⁰ 01′ 53 E	20.57	9.56	0.46	13.94	10.93	0.25
RES 2	15⁰ 48′ 13 S 35⁰ 01′ 27 E	13.84	8.64	0.43	13.53	8.47	0.26
RES 3	15⁰ 49′ 01 S 35⁰ 00′ 03 E	22.94	7.47	0.34	12.64	7.85	0.54
CBD 1	15⁰ 47′ 13 S 35⁰ 00′ 41 E	22.57	10.85	0.35	21.46	11.69	0.53
CBD 2	15⁰ 47′ 18 S 35⁰ 00′ 20 E	13.64	9.46	0.41	18.45	9.95	0.28
CBD 3	15⁰ 47′ 12 S 35⁰ 00′ 11 E	23.94	11.04	0.37	14.53	11.56	0.38

concentration were recorded in Hospitals and low dense residential areas during of low wind speeds (<5 m s^{-1}). This might also be as a result of less exposure to emissions from domestic fuel combustion, less traffic and good vegetative cover within these areas. The higher values were recorded in open markets and schools. At market site 1 and site 2, the mean $PM_{2.5}$ concentrations were 31.8 μg/m³, and 42.5 μg/m³ respectively. These observed $PM_{2.5}$ concentrations exceeded the 24-h upper limit set by Malawi Bureau of Standards on air quality. The observed exceedance at site 1 and 2 could be attributed to the direct exposure of emissions from dirty solid fuel combustion and dusty storms observed during the $PM_{2.5}$ monitoring period. The $PM_{2.5}$ mass concentrations in CBD were midway between that of schools and industries.

4 Conclusion

In this study, $PM_{2.5}$ pollution levels in Blantyre City, Malawi, were measured and then compared with 24-houly safety limits set by MBS and WHO. The main finding of this $PM_{2.5}$ monitoring exercise was that $PM_{2.5}$ in Blantyre were above safe limits for both local and WHO air quality standards on particulate pollution of 8 μg/m³ and 25 μg/m³ for $PM_{2.5}$ 24-h period respectively. About 65% of the monitored sites had $PM_{2.5}$ values exceeding this $PM_{2.5}$ safe limit of WHO. This shows that there are significant air quality problems emanating from high levels of $PM_{2.5}$ concentrations in the industrial areas, markets and school areas within Blantyre. Although $PM_{2.5}$ concentration levels were lower in some locations (in reference to WHO guidelines), it is important that the duration of exposure and source characterization of pollutants be studied further with an extended period. MBS and WHO compliance of $PM_{2.5}$ concentrations for 2 out of 6 monitoring sets points cannot confirm that air quality is safe in the City of Blantyre. For this reason, this research recommends long term $PM_{2.5}$ levels and mobility monitoring and assessment in Malawi.

Acknowledgements The authors would like to thank Malawi University of Business and Applied Sciences, and Environmental Solution Consultants for supporting this study and providing air monitoring equipment (Dylos DC1100 PRO Laser Particle Counter) respectively.

References

Cohen, J., H. Ross A., Bart O., Kiran Dev P., Michal K., Nino K., Kersten G., C. Arden P., Isabelle R., and Kirk R. S. (2024). Urban Air Pollution. Comparative Quantification of Health Risks: Global and Regional Burden of Disease Attributable to Selected Major Risk Factors. World Health Organization, 2004, pp. 1353–434. JSTOR, http://www.jstor.org/stable/resrep27829.22.

Derafshi, M., Black, C., & Agnew, R. (2017). Pesticide knowledge, attitudes, and clothing practices of Turkish Farmers. *Environmental Management and Sustainable Development, 6*(1), 149.

Jary, H. et al. (2015). Study protocol : The effects of air pollution exposure and chronic respiratory disease on pneumonia risk in urban Malawian adults - the Acute Infection of the Respiratory Tract Study (The AIR Study). *BMC Pulmonary Medicine,* 1–8.

Mapoma, H. W. T., Xie, X., Wilson, H., Mapoma, T., & Xie, X. (2023). State of air quality in Malawi. *Journal of Environmental Protection (Irvine,. Calif), 2013*(November), 1–7.

MBS. (2017). Malawi Bureau of standards promoting standardization and quality assurance in Malawi catalogue of Malawi standards 2017 Malawi Bureau of Standards Moirs Road P.O Box 946 Blantyre MALAWI E-mail : mbs@mbsmw.org, website : www.mbsmw.org.

Organizaci6n Mundial de la Salud (OMS). (2021). WHO global air quality guidelines. *Coastal and Estuarine Processes,* 1–360.

Obstructive, C., Disease, P., Veterans, C. C., & Markers, R. (2018). Chronic Obstructive pulmonary disease and herbicide exposure in Vietnam-Era U.S. Army Chemical Corps Veterans. *15*(April), 1–2.

Okello, G., et al. (2023). Air quality management strategies in Africa: A scoping review of the content, context, co-benefits and unintended consequences. *Environment International, 171*(October 2022), 107709.

Osipov, S., et al. (2022). Severe atmospheric pollution in the Middle East is attributable to anthropogenic sources. *Communications Earth & Environment, 3*(1), 1–10.

Thulu, F., et al. (2023). Electromagnetic frequency pollution in Malawi: A case of electric field and magnetic flux density pollution in Southern Africa.

Thurlow, J., Zhu, T., & Diao, X. (2012). Current climate variability and future climate change: Estimated growth and poverty impacts for Zambia. *Review of Development Economics, 16*(3), 394–411.

Xu, H., et al. (2020). Personal exposure to $PM_{2.5}$ emitted from typical anthropogenic sources in southern West Africa: Chemical characteristics and associated health risks to cite this version: HAL Id : hal-03041633 personal exposure to $PM_{2.5}$ emitted from typical anthropo.

Turbulent Kinetic Energy and Turbulence Dispersion Characteristics Over a Mixed Crop Area in Benin

Ossénatou Mamadou, Miriam Hounsinou and Basile Kounouhéwa

Abstract

In this study, we analyze turbulent kinetic energy and turbulent dispersion characteristics of four 15-day periods selected in the year 2008. Data were acquired using the eddy covariance method above a mixed crop area located in the northern Benin. The Monin Obukhov Similarity (MOS) theory was used to examine the relationship between the turbulent dispersion characteristics and stability parameters for the four periods investigated. It has been found that the turbulent kinetic energy, friction velocity and turbulent dispersion characteristics showed distinctive variations during the selected periods. During unstable atmospheric conditions, the normalized ratios of lateral, horizontal and vertical wind speeds are larger with respect to those obtained in stable stratification of the atmosphere. The results demonstrate also that the normalized standard deviations of wind speed components follow MOS theory under unstable atmospheric conditions.

Keywords

Turbulent kinetic energy · Turbulent dispersion characteristics · Normalized standard deviations · Monin Obukhov Similarity

O. Mamadou (✉) · M. Hounsinou
Institute of Mathematics and Physics, BP 613, Dangbo, Bénin
e-mail: ossenatou.mamadou@imsp-uac.org

O. Mamadou · M. Hounsinou · B. Kounouhéwa
LPR, FAST/UAC, Abomey-Calavi, Bénin

1 Introduction

In the Atmospheric Boundary Layer (ABL), exchanges of heat, momentum, moisture, atmospheric gases occur mainly through turbulence which is the principal physical process in this part of the atmosphere. Indeed, turbulence plays a crucial role by favoring the efficient mixing of atmospheric particles, their accumulation and transportation from surface to upper atmosphere thus the genesis of various weather phenomena (Prasad et al., 2019) and turbulent diffusion of air pollutants. Its understanding remains therefore useful to evaluate atmospheric models as well as for issues related to the dispersion of pollutant (Cohan et al., 2011; Kesarkar et al., 2007). It is noteworthy that, most of the air quality dispersion models require turbulence data to the proper simulation of the atmospheric dispersion of air pollutants. Those data considered most significant are the turbulence intensity, stability parameter, friction velocity, sensible heat, Monin Obukhov length (Seaton et al., 2022; Liu et al., 2021; Yuval et al., 2020). A recent study conducted by (Coccia, 2021) has shown the link between atmospheric stability, the degree of air pollution and the diffusion of COVID-19 in the atmosphere. However, the boundary layer meteorological factors vary within the day and also according to the geographical locations which in turn alter the turbulence characteristics. As a result, it becomes difficult to use these turbulence characteristics estimated over a given location for another one. In a recent work conducted by Hounsinou et al. (2022) above a woodland forest in the northern region of Benin, authors demonstrated a mismatch between their data-driven models established for horizontal and vertical turbulent characteristics and those available in literature. Similar results have been obtained when comparing carbon dioxide turbulent characteristics from a forest and those obtained above a cultivated site in the same climatic region (Hounsinou et al., 2023). These findings emphasize the need to explore

K. W. Fomba et al. (eds.), *Advances in Air Quality Research in Africa*,
Advances in Science, Technology & Innovation, https://doi.org/10.1007/978-3-031-53525-3_18

the boundary layer turbulence dispersion characteristics in order to deliver more insights for this area.

This work is partly addressing this need using one-year long observation with a focus on four 15-day periods representative of the corresponding season, at a mixed crop site in Sudanian climate Northern Benin, with the Monin and Obhukov Similarity (MOS) theory. Formalized in a rigorous framework, the theory of similarity in flat and homogeneous areas, that is the MOS theory should be verified whether this is applicable for non-homogenous areas such as mixed crops. We hypothesize that the meteorological conditions in each phase of crop development thus each 15-day periods are significantly distinct from others. Turbulence dispersion characteristics and the similarity relations, which are affected by the boundary layer meteorological conditions would be changed during different phases, thus this has to be fully examined. After the introduction, we describe the materials and method used in this study. The results obtained are presented in Sect. 3 followed by a short discussion. Finally, the conclusion is given in Sect. 4.

2 Materials and Method

2.1 Study Site Description

The study location is in Nalohou (latitude 9.74°N, longitude 1.60°E, a.s.l of 449 m), a village in northern Benin. It lies approximately 10 km north of Djougou, Donga department and represents a typical area under Sudanian climate. In this region, we have an alternate of wet and dry seasons following by two transitional phases (Mamadou et al., 2014) with a mean rainfall amount of 1200 mm. The observation site is installed above the herbaceous fallow surrounded by mixed crops and trees from north to south. Toward the south of the measurement tower, the vegetation is dominated by cultivated areas (cassava, yam, maize, and groundnut). The herbaceous layer under the scope of this study has grown in 2008 between April (0 m) to October (2.5 m); dry herbs have been burnt on the 29th of October (Mamadou et al., 2014).

2.2 Meteorological and Turbulence Observations

Meteorological data were collected automatically at a 1 min resolution and averaged at 15 min, including wind speed and direction, air temperature and humidity, four components of the solar radiation. The 20 Hz turbulence data, including high frequency wind velocity and sonic temperature, carbon dioxide, pressure and absolute humidity data were collected with the eddy covariance sensors at the height of 4.95 m. Details regarding the study site and observations can be found in Mamadou et al. (2014). The quality control methods and correction have been applied to turbulent data to discard problematic measurements prior their analysis.

2.3 Method

We analyzed in this study the friction velocity (u_*), atmospheric stability ($\xi = z - d/\Lambda$), turbulent kinetic energy (TKE) and the three wind speed components (u, v, w) normalized by u_*. The later are functions of (ξ) according to Monin–Obukhov's Similarity theory and expressed as:

$$\sigma_i/u_* = f_i((z-d)/\Lambda) = a_i(1 + b_i\xi)_i^c, \tag{1}$$

where σ_i ($i = u$, v, w) represents the standard deviations of the wind speed components and the atmospheric scalars; d (m) and z (m) the displacement and measurement heights respectively; f_i (.) the similarity function relatively to (ξ) and (Λ) the Obhukov length (Foken, 2017). The coefficients a_i, b_i and c_i in (Eq. 1) are the fitting parameters. The turbulent kinetic energy is obtained as:

$$\text{TKE} = \left(\sigma_u^2 + \sigma_v^2 + \sigma_w^2\right)/2. \tag{2}$$

3 Results and Discussion

Figure 1 displays the diurnal variations of wind speed (Ws), friction velocity (u_*), turbulent kinetic energy (TKE), and three wind speed components normalized by friction velocity for each 15-day period: P1 (January 18th–February 01st); P2 (February 26–March 11th); P3 (July 07th–July 21st) and P4 (October 24th–November 07th) in the year 2008. These periods have been chosen due to their particularity regarding both atmospheric and surface conditions already studied extensively in Mamadou et al. (2014).

The (P1) period is situated in dry season characterized by the presence of aerosol dusts which lowered the incoming shortwave radiation; the (P2) lies in the transition between dry and wet season during which first rainy events that occurred in the region; the (P3) period is located in the wet season with a lot of clouds while the last period (P4) is situated between the wet and dry season. At this time of the year, the crops are already harvested. As expected, the diurnal cycle of all turbulent dispersion characteristics shows different patterns in all 15-day periods analyzed. During daytime, when turbulence is well developed (u_* higher) and the atmosphere is unstable ($\xi < 0$), relatively strong wind speeds (~1.5 < Ws < ~5 m/s) depending on the period prevail

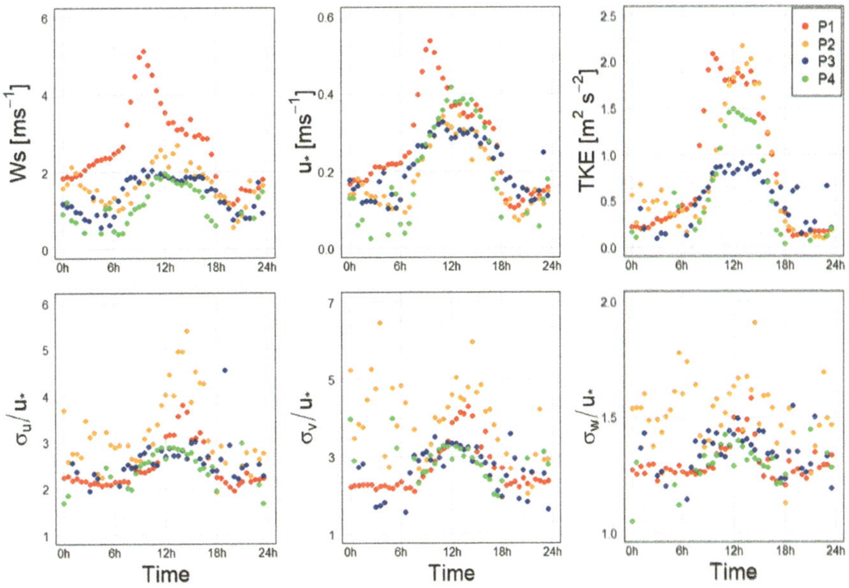

Fig. 1 Diurnal cycle of wind speed (W_s), friction velocity (u_*), turbulent kinetic energy (TKE) and the normalized ratio of *u*, *v* and *w* for the selected periods: red (P1), orange (P2), blue (P3) and green (P4) of the 2008 year

at the site, and the normalized ratios σ_u/u_*, σ_v/u_* and σ_w/u_* are larger with respect to those obtained at night especially in P1 and P3. Some scattering was observed, however, in the nighttime values of σ_v/u_* and σ_w/u_* during P2 and P4.

The turbulent standard deviations in P1 (days with dusts) and P2 are evidently higher on average than those of P3 and P4 (Fig. 1). It has been found that u_*, Ws and TKE in all periods increase with time and show maximum value around midday and thereafter they decrease (Fig. 1). Slight time lag between thermal effects starts dominating during daytime thereby giving an early maximum for TKE and u_* in P1. Another interesting feature that come out from this comparison may also be noted here is that turbulence characteristics values are higher in the early hours of the day than those at the noon except for v and w in the moistening period.

The relationships between the normalized wind speed components and the stability parameter for each period are shown in Fig. 2. Note that the coefficient a_i and b_i vary according to periods because the normalized ratios do not have the same distribution during these periods (Hounsinou et al., 2022). These relationships are in the form of Eq. (1)

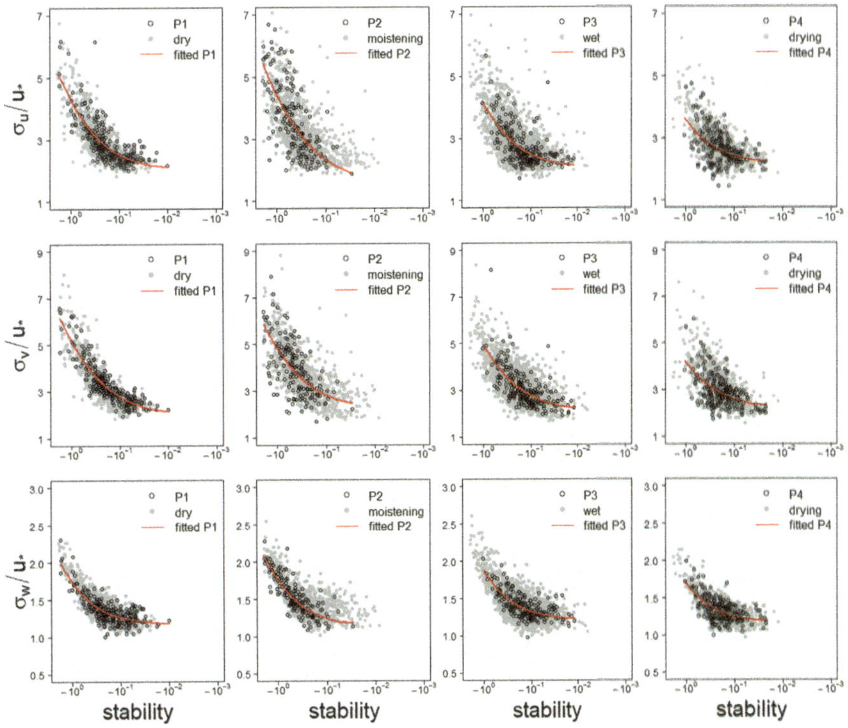

Fig. 2 Relationship between turbulent standard deviations of u, v and w normalized by u_* and the stability parameters during unstable conditions in P1, P2, P3 and P4 (black dots) and the gray color represents the whole dry, moistening, wet and drying seasons

with c_i equal to 1/3 as suggested by MOS theory. This demonstrates that turbulent characteristics obey to MOS theory especially in unstable conditions.

4 Conclusion

This work attempts to estimate turbulent parameters required by some dispersion models. It also provides insight into the role of atmospheric stability in determining turbulent characteristics. The result obtained could be beneficial for better representation of turbulent parameterizations in dispersion models over the region. The nonlinear relationship between these parameters as a function of stability we studied cannot be used at this stage for forecasting purposes. Further work which could be based on the turbulence measurements taken under various stability conditions, are necessary to extend the applicability of the results.

Acknowledgements This work was carried out with the aid of a grant from UNESCO and the International Development Research Centre, Ottawa, Canada. The views expressed herein do not necessarily represent those of UNESCO, IRDC or its Board of Governors.

References

Coccia, M. (2021). The effects of atmospheric stability with low wind speed and of air pollution on the accelerated transmission dynamics of COVID-19. *International Journal of Environmental Studies, 78*(1), 1–27.

Cohan, A., Wu, J., & Dabdub, D. (2011). High-resolution pollutant transport in the San Pedro Bay of California. *Atmospheric. Pollution. Research., 2*, 237–246.

Hounsinou, M., Mamadou, O., Wudba, M., Kounouhewa, B., & Cohard, J.-M. (2022). Integral turbulence characteristics over a clear woodland forest in northern Benin (West Africa). *Atmospheric Research, 268*, 105985.

Hounsinou, M., Mamadou, O., & Kounouhewa, B. (2023). Turbulence characteristics in the atmospheric surface layer over a heterogeneous cultivated surface in a tropical region. *Boundary Layer Meteorology, 188*, 321–344.

Kesarkar, A. P., Dalvi, M., Kaginalkar, A., & Ojha, A. (2007). Coupling of the weather research and forecasting model with AERMOD for pollutant dispersion modeling. A case study for PM_{10} dispersion over Pune, India. *Atmospheric Environment, 41*(9), 1976–88.

Liu, L., Shi, Y., & Hu, F. (2021). Characteristics and similarity relations of turbulence dispersion parameters under heavy haze conditions. *Atmospheric Pollution Research, 12*(2), 330–40.

Mamadou, O., Cohard, J. M., Galle, S., Awanou, C. N., Diedhiou, A., Kounouhewa, B., & Peugeot, C. (2014). Energy fluxes and surface characteristics over a cultivated area in Benin: Daily and seasonal dynamics. *Hydrology Earth Systems Science, 18*, 893–914.

Prasad, K. B., Hari, R. R., Srinivas, C. V., Bagavth Singh, A., Naidu, C. V., Baskaran, R., & Venkatraman, B. (2019). Turbulence Characteristics of Surface Boundary Layer over the Kalpakkam Tropical Coastal Station, India. *Meteorology and Atmospheric Physics, 131*(4), 827–43.

Seaton, M., O'Neill, J., Bien, B., Hood, C., Jackson, M., Jackson, R., Johnson, K., et al. (2022). A multi-model air quality system for health research: road model development and evaluation. *Environmental Modelling & Software, 155*, 105455.

Yuval, T. T., Raanan, R., Yoav, L., Levy, I., & Broday, D. M. (2020). Emissions vs. turbulence and atmospheric stability: A study of their relative importance in determining air pollutant concentrations. *Science of The Total Environment, 733*, 139300.

Air Pollution and Health Effects

Indoor Air Quality: Assessment of Particulate Matter and Non-cancerous Inhalation Health Risk in Nigeria

Francis Olawale Abulude, Samuel Dare Oluwagbayide, Akinyinka Akinnusotu, Kikelomo Mabinuola Arifalo, Sunday Acha, Amoke Monisola Kenni and Ademola Adamu

Abstract

The assessment of global indoor air quality (IAQ) is important because of the threat (health risks) posed by pollutants released into the house as a result of various indoor and outdoor activities. There haven't been many studies on IAQ and health, so particulate matter (PM_1, $PM_{2.5}$, and PM_{10}) and its attendant health risk on infants, children, and adults was studied in Akure, Ondo State, Nigeria, using standard USEPA methods. The results show that PM_1 ranges between 0.0 and 237.0 µg/m^3, while $PM_{2.5}$ and PM_{10} have mean ± SD of 17.29 ± 21.76 and 18.89 ± 27.15 respectively. $PM_{2.5}$ and PM_{10} air quality index (AQI) have mean values of 43.21 and 31.48, respectively. The mean acute exposure values for the infant, child, and adult are 0.43, 0.44, and 0.27, respectively. The risk Hazard Quotient (HQ) values are < 1. An HQ of < 1.0 indicates a negligible risk, which means that the PM evaluated in this study is unlikely to pose a health risk to a sensitive individual. In conclusion, the results show that some of the PM values exceed the WHO 24 h annual mean maximum exposure limit as well as the Nigerian National Ambient Air Quality Standard. Although there are times when the air quality is hazardous, it is heartening to note that the AQI is generally good during the study periods. On an acute basis, the indoor assessment is unlikely to cause adverse health effects in infants, children, or adults.

Keywords

Particulate matter · Air quality index · Health risk · Hazard quotient · WHO

1 Introduction

Air pollution is a major issue that affects numerous countries. Poor air quality not only causes and exacerbates a range of health problems, resulting in seven million premature deaths each year, but it also harms the environment, climate change, and severe weather events (WHO, 2021). Airborne particles, particularly fine particles, have been linked to a variety of health issues, including increased illness and hospital readmissions (Abulude et al., 2018; Habeebullaha et al., 2017). To address these lingering issues, AQI understanding and use are required. The purpose of the AQI is to allow citizens to see how good their environment is. It helps us avoid pollution hotspots and organize their daily movement patterns to minimize exposure to potentially harmful pollutants by observing at a glance. According to Opara et al. (Opara et al., 2021), risk and vulnerability assessment is a technique that analyses the extent of inhalation to PM pollutants in conjunction with the life cycle impact assessment (LCIA) method. It

F. O. Abulude (✉)
Environmental and Sustainable Research Group, Science and Education Development Institute, Akure, Ondo State, Nigeria
e-mail: walefut@gmail.com

S. D. Oluwagbayide
Department of Agricultural and Bio-Environmental Engineering, Federal Polytechnic, P.M.B. 50, Ilaro 112106, Ogun State, Nigeria

A. Akinnusotu
Department Science Laboratory Technology, Rufus Giwa Polytechnic, P.M.B. 1019, Owo, Ondo State, Nigeria

K. M. Arifalo · A. Adamu
Department of Chemical Sciences, Bamidele Olumilua University of Education, Science, and Technology, Ikere-Ekiti, Ekiti State, Nigeria

S. Acha
Department of Geography, Federal University of Technology, Minna, Niger State, Nigeria

A. M. Kenni
Department of Science Education, Bamidele Olumilua University of Education, Science, and Technology, Ikere-Ikere, Ekiti State, Nigeria

K. W. Fomba et al. (eds.), *Advances in Air Quality Research in Africa*,
Advances in Science, Technology & Innovation, https://doi.org/10.1007/978-3-031-53525-3_19

entails the use of a building users scenario that is based on pollutant dosages and usually further evaluates the risk to human health for a given period of time. Many scientists have studied the health consequences of PM and found that it can induce respiratory problems, rhinosinusitis, cardiovascular diseases, asthma, and chronic obstructive pulmonary disease (Mohammed et al., 2015). The goal of this work was to concentrate on indoor particulate matter ($PM_{2.5}$ and PM_{10}), AQI, and non-cancerous inhalation health risk assessments in Nigeria.

2 Methodology

Nigeria is a highly populated (206.1 million people) African country with 36 states and a federal capital territory that is rapidly growing in terms of population, urbanization, industrialization, commercialization, and vehicular activity, to name a few. Ondo State is one of the states. Akure is the administrative headquarters of Ondo State, one of several oil-rich Niger-Delta in Nigeria's southwest geopolitical zone. It is situated 700 km southwest of Abuja and 311 km north of Lagos, Nigeria (Akinwumiju et al., 2021). The study residence is in Akure, the capital city of Ondo State in Southwestern Nigeria.

ENEA-Italian National Agency for New Technologies, Energy, and Environment, Department of Sustainable Development, Brindisi Research Center, SS. 7, Appia, 706 km, 72,100 Brindisi, Italy (Suriano et al., 2020) designed and developed the sensor used in this study. The study lasted 30 days (01/06/2021–30/06/2021) and took place in one of the building's rooms (Fig. 1). The sensor was suspended from a stand 1.5 m above the ground level. The monitoring methods were carried out in accordance with the manufacturer's recommendations (Suriano, 2021). The data collected during the monitoring was saved in the SD card embedded in the sensor, downloaded, and statistically analyzed using Minitab 19.2020.1 and Excel 2013.

3 Results and Discussion

The mean concentrations of indoor particulate matter (PM_1, $PM_{2.5}$, and PM_{10}) in Table 1 are 11.86 ± 10.61, 17.29 ± 21.76, and 18.29 ± 27.15 µg/m^3, respectively. These are significantly higher than the annual WHO (2021) limits but significantly lower than the Nigerian National Ambient Air Quality Standards ($PM_{2.5}$—35 µg/m^3 and PM_{10}—150 µg/m^3) (ESREA, 2021). The levels obtained are the result of indoor activities such as cooking, frying, insecticides (mosquito coil, cigarette smoke, sweeping), poor ventilation, and outdoor activities such as dust particle re-suspension.

The AQI (Table 1) shows mean levels of 43.21 ± 54.00 ($PM_{2.5}$) and 31.48 ± 45.25 (PM_{10}). The AQI of PM_1 is not included due to the lack of a standard for the PM. The AQI's minimum and maximum values are 0.00–1760.00 ($PM_{2.5}$) and 0.00–1608.33 (PM_{10}), respectively. This suggests that if the maximums are reached, some vulnerable people may face health risks.

Table 2 displays the health risk (non-carcinogenic) evaluations for acute particulate matter exposure. Infant (0.430.56), Child (0.440.56), and Adult (0.27036) are the

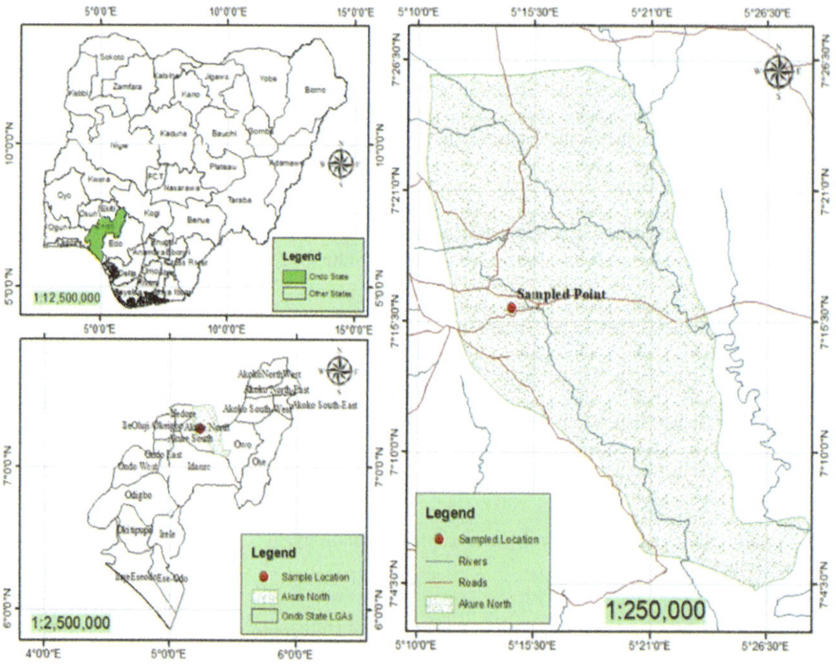

Fig. 1 The map showing the location used for the study

Table 1 Particulate matter and air quality index

	PM			AQI		
Statistics	PM_1	$PM_{2.5}$	PM_{10}	PM_1	$PM_{2.5}$	PM_{10}
Mean	11.86	17.29	18.89	–	43.21	31.48
StDev	10.61	21.76	27.15	–	54.00	45.25
Minimum	0.00	0.00	0.00	–	0.00	0.00
Q1	6.00	9.00	9.00	–	22.50	15.00
Q3	16.00	21.00	23.00	–	52.52	38.33
Maximum	237.00	704.00	965.00	–	1760.00	1608.33
Skewness	9.11	17.76	1.84	–	17.76	19.84
Kurtosis	151.20	445.13	537.47	–	445.13	537.47

PM—Particulate Matter, AQI—Air Quality Index, StDev—Standard Deviation

Table 2 Non-cancerous inhalation health risk

Non-carcinogenic pollutants exposure						
	Acute exposure			Hazard quotient[a]		
	Infant	Child	Adult	Infant	Child	Adult
PM_1	0.32 ± 0.28	0.31 ± 0.28	0.20 ± 0.18	–	–	–
$PM_{2.5}$	0.46 ± 0.58	0.45 ± 0.59	0.29 ± 0.37	0.5 ± 0.07	0.05 ± 0.07	0.03 ± 0.04
PM_{10}	0.05 ± 0.71	0.50 ±s 0.72	0.32 ± 0.45	0.03 ± 0.03	0.03 ± 0.04	0.02 ± 0.02
Overall						
Means	0.43	0.44	0.27	0.04	0.06	0.02
StDevs	0.56	0.56	0.36	0.05	0.05	0.03

[a] There is no reference exposure level for $PM_{1.0}$ because of this, the relevant exposure threshold has no evidence of the toxicological harmful effect on human health

overall results. When considering infants, children, and adults on an acute basis, the results show that PM_{10} has the highests hazardous potential.

The infant, child, and adult have mean Hazard Quotient (HQ) exposure values of 0.04, 0.06, and 0.02, respectively. The (HQ) values are<1. A HQ of<1.0 implying a negligible risk, indicating that the PM investigated in this study is unlikely to pose a health risk to a sensitive individual.

4 Conclusion

The PM levels examined in this research are higher than the annual WHO limits but lower than the Nigerian National Ambient Air Quality Standards. The AQI obtained at low levels indicates that people living within the study area are safe for outdoor and indoor activities, but the maximum levels indicate that some vulnerable people may face health risks. When considering infants, children, and adults on an acute basis, the results show that PM10 has the highest hazardous potential. Since the calue of HQ was less than 1, the PM evaluated in this study is unlikely to pose a health risk to a sensitive individual. To avoid pollution, efforts should be directed toward constant monitoring of the IAQ.

Acknowledgements Dr. Suriano and Dr. Penza of ENEA-Italian National Agency for New Technologies, Energy, Environment. Sustainable Development Department, Research Center of Brindisi, SS. 7, Appia, 706 km, 72100 Brindisi, Italy provided the SentinAir sensor used in this study. The authors would also like to thank Emmanuel Awogbindin of the Department of Computer Science at the Federal College of Agriculture in Akure, Ondo State, Nigeria, for the statistical analyses.

References

Abulude, F. O., Ndamitso, M. M., Iyaka, Y. A., & Abdulkadir, A. (2018). Study of particulate matter of Akure, Nigeria using a sharp-cut inertial filter combined with an impactor-A preliminary study. *Journal of Atmospheric and Solar-Terrestrial Physics, 179*, 396–404.

Akinwumiju, A. S., Ajisafe, T., & Adelodun, A. A. (2021). Airborne particulate matter pollution in Akure Metro City, Southwestern Nigeria, West Africa: Attribution and meteorological Influence. *Journal of Geovisualization and Spatial Analysis, 5*, 11.

Habeebullaha, T. M., Mohammeda, A. M. F., Attalaa, O. A., Munira, S., & Alobaidi, B. S. (2017). Health impact of indoor PM_{10} in Makkah, KSA. In *IAQHAC 2017 – Workshop on Indoor Air Quality in Hot Arid Climate, Kuwait*, April 3–4.

Mohammed, A. M. F., Munir, S., & Habeebullah T. M. (2015). Characterization of atmospheric aerosols in Makkah. *International Journal of Agricultural and Environmental Researchs* (December), 1–17.

Opara, A. I., Akaolisa, C. Z., Akakuru, C. O., Nkwoada, A. U., Ibe, F. C., Verla, A. W., & I. C. Chukwuemeka. (2021). Particulate matter exposure and non-cancerous inhalation health risk assessment of major dumpsites of Owerri metropolis, Nigeria. *Environmental Analysis Health and Toxicology*, *36*(4), e202102 5 (2021).

Environmental Standards and Regulations Enforcement Agency (ESREA) (Establishment) (Ammended) Act, 2018. National Environmental (Air Quality Control) (2021).

Suriano, D., Cassano, G., & Penza, M. (2020). Design and development of a flexible, plug-and-play, cost effective tool for on-field evaluation of gas sensors. *Journal of Sensors*. https://doi.org/10.1155/2020/8812025

Suriano, D. (2021). A flexible tool for data acquisition from heterogeneous low-cost gas sensors and other devices. GitHub - domenico-suriano/SentinAir.

WHO. (2021). Ambient (outdoor) air pollution. WHO Air quality guideline values. Retrieved June 2, 2022, from https://www.who.int/news-room/fact-sheets/detail/ambient-(outdoor)-air-quality-and-health.

Air Quality Monitoring Assists Meeting the Sustainable Development Goals in Ethiopia

Araya Asfaw, Christina Isaxon, Ebba Malmqvist, Sina Hasheminassab and David J. Diner

Abstract

Air pollution is a global challenge impacting the lives of millions of people, especially in low-income countries. Improving air quality is linked to several of the United Nations Sustainable Development Goals (SDGs). The state of particulate matter (PM) pollution in Ethiopia and the importance of a national ambient air quality monitoring program are discussed. The contributions of NASA's upcoming satellite-based Multi-Angle Imager for Aerosols (MAIA) investigation to the SDGs are highlighted.

Keywords

Air quality · Particulate matter · UN sustainable development goals · MAIA

1 Introduction

Air pollution has been recognized as a global issue for more than a century. According to the Global Burden of Disease study (GBD, 2020), air pollution is the largest environmental risk factor for disease and premature deaths around the world. The World Health Organization (WHO) attributes seven million preventable deaths annually to the joint effects of household and ambient air pollution (about 2.5 times more than from AIDS, tuberculosis, and malaria combined), with nearly the entire global population (99%) breathing air containing pollutants exceeding the limits set by WHO guidelines (Web-1). About 91% of deaths from outdoor air pollution alone occur in low and middle income countries (Web-2).

Particulate matter (PM) consists of airborne solid, liquid, organic, and inorganic particles of different sizes. Constituents include sulfates, nitrates, ammonium compounds, elemental and organic carbon, and dust. Particles ≤ 10 µm in diameter (PM_{10}) are readily inhalable and those ≤ 2.5 µm in diameter ($PM_{2.5}$) can penetrate deep inside the lungs. Short- and long-term exposures to such particles, particularly those having smaller sizes, have severe health impacts, including various types of noncommunicable diseases, such as stroke, heart disease, and lung cancers among adults. Children under five years of age are highly susceptible to the effects of unhealthy air. Based on analysis of air pollution in 117 countries, the WHO guidelines for PM exposure were updated in 2021. For $PM_{2.5}$ (PM_{10}), the new mean exposure recommendations are ≤ 5 µg/m^3 (15 µg/m^3) annually and ≤ 15 µg/m^3 (45 µg/m^3) daily.

A. Asfaw (✉)
Institute of Geophysics, Space Science and Astronomy, Addis Ababa University, Addis Ababa, Ethiopia
e-mail: araya.asfaw@aau.edu.et

C. Isaxon · E. Malmqvist
Lund University, Lund, Sweden
e-mail: Christina.isaxon@design.lth.se

E. Malmqvist
e-mail: Ebba.malmqvist@med.lu.se

S. Hasheminassab · D. J. Diner
Jet Propulsion Laboratory, Pasadena, CA, USA
e-mail: sina.hasheminassab@jpl.nasa.gov

D. J. Diner
e-mail: David.J.Diner@jpl.nasa.gov

K. W. Fomba et al. (eds.), *Advances in Air Quality Research in Africa*,
Advances in Science, Technology & Innovation, https://doi.org/10.1007/978-3-031-53525-3_20

2 Linkages Between Air Quality in Africa and United Nations Sustainable Development Goals

In Africa, more than a million deaths annually are associated with air pollution (Fisher et al., 2021). Indoor air pollution due to burning unclean fuels such as wood, charcoal, and dung with inefficient cookstoves in unventilated environments accounts for about 64% of these deaths. The remaining 36% are linked to outdoor air pollution, primarily driven by vehicular emissions of particulates.

Worldwide, 759 million people do not have access to electricity; 78% of them live in Africa. According to the World Bank, more than 2.6 billion people do not have access to clean fuels and appropriate stoves, of which 910 million live in sub-Saharan Africa. For seven countries in Africa, including Ethiopia, less than 5% of the population has access to clean cooking. By 2050, the population of Africa is expected to double. Two-thirds of people will likely reside in urban areas. With population gains outpacing access to clean and sustainable energy, air pollution can be expected to present a growing concern (Web-3).

Development of strategies to overcome the shocks and chronic stresses associated with rapid growth, including increased environmental pollution, is essential. Improving air quality in Africa is directly aligned with several United Nations Sustainable Development Goals, or SDGs (Web-4). Goal 3, for example, aims at improving health, well-being, and safety. Reducing air pollution is necessary to meet this goal, and one way of achieving it is by providing clean and affordable energy (Goal 7). Since women and young children are strongly impacted by indoor air pollution, improving air quality is linked to gender equality (Goal 5). Goal 11 aims at making cities and human settlements inclusive, safe, resilient, and sustainable. Continuous air quality monitoring is essential for measuring progress towards attaining these goals.

3 State of Air Pollution in Ethiopia

Ethiopia is the second most populous country in Africa with a population in excess of 100 million people. Since the 1950s the population has quadrupled. It is a low-income country with less than $1,000 Gross Domestic Product per capita. Energy consumption, 20 gigajoules per person per year, is one of the lowest in Africa. Traditional fuels (wood, charcoal, and dung) provide more than 90% of the primary energy. About 7% of the energy needs are met by imported oil and petroleum products, and electricity from renewable sources (mainly hydropower) accounts for the remainder (Web-5).

In Addis Ababa, the capital city of Ethiopia, a study of ambient air quality from 2017 to 2021 showed that the daily mean concentration of $PM_{2.5}$ exceeded 42 $\mu g/m^3$, nearly three times the WHO guideline, with maximum values exceeding 100 $\mu g/m^3$ (Kumie et al., 2021). Tefera et al. (2016) reported values as high as 280 $\mu g/m^3$. Indoor $PM_{2.5}$ exposure can reach values several times larger (Sanbata et al., 2014).

Air quality monitoring is in its early stages. The limited number of studies to date motivates the need for continuous air quality measurements with greater spatial and temporal coverage and obtaining a deeper understanding of the human health impacts. The joint use of satellite data, surface-based measurements, and chemical transport model outputs provides a means of accomplishing this objective.

4 The MAIA Investigation

In 2016, NASA selected the Multi-Angle Imager for Aerosols (MAIA) satellite investigation (Diner et al., 2018). MAIA's primary objective is to associate exposure to different sizes and compositional mixtures of airborne PM with adverse health outcomes. A set of 11 Primary Target Areas (PTAs) around the world has been identified for conducting this investigation. PTAs are 360 km × 480 km regions that include major population centers covering a range of PM concentrations and types. Two PTAs are located in Africa: Central Ethiopia, including Addis Ababa and the nearby city of Adama; and Gauteng Province of South Africa, including Johannesburg, Soweto, and Tshwane (Fig. 1).

4.1 Satellite Instrument

The MAIA instrument is designed to provide sensitivity to aerosol concentration, size, shape, and absorption by observing sunlight scattered by the Earth's atmosphere and surface. A pointable camera acquires multiangular imagery in 14 spectral bands spanning the ultraviolet, visible/near-infrared, and shortwave infrared regions of the electromagnetic spectrum. Three of the bands are polarimetric. The instrument is currently undergoing integration and test at the Jet Propulsion Laboratory and will be launched into Earth orbit later this decade. The baseline mission duration is 3 years.

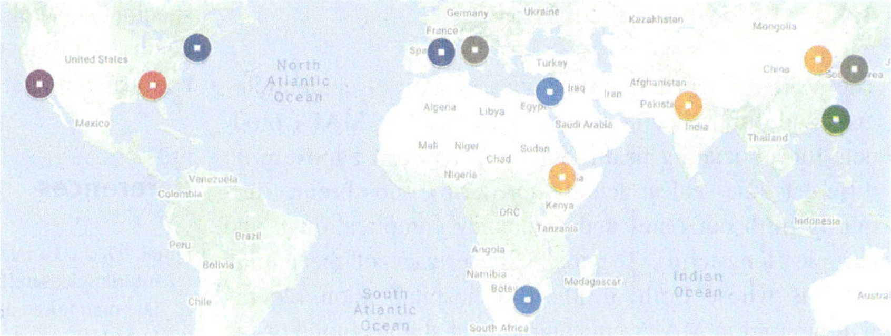

Fig. 1 MAIA Primary Target Areas

4.2 Surface Monitor Network and Other Data Sources

A set of geostatistical regression models (GRMs) will be used to transform the satellite-based aerosol retrievals into near-surface PM concentrations. Surface monitors are used to calibrate the GRMs. The paucity of PM measurements in Ethiopia, especially from instruments sensitive to particle speciation, has required the deployment of a variety of sensors. The US Agency for International Development (USAID) is funding surface monitor operation by Lund University and the Ethiopian Wildlife and Natural History Society as well as chemical analysis of filter-based samples by Washington University (WashU), Colorado State University (CSU), and University of California–Davis. Overseas shipments of the instruments and consumables have been facilitated by the US Department of State (USDOS), which also hosts some of the instruments at embassy sites (Web-6). MAIA utilizes the following surface PM monitors in Ethiopia:

(1) An AirPhoton SS5i-$PM_{2.5}$ filter-based sampling station as part of the international Surface Particulate Matter Network (SPARTAN) managed by WashU;
(2) Two CSU Aerosol Mass and Optical Depth (AMOD) samplers;
(3) Two AethLabs MA350 black carbon aethalometers;
(4) A network of PurpleAir-II-SD low-cost total PM sensors;
(5) A CIMEL aerosol sunphotometer provided by the NASA Goddard Space Flight Center as part of the international Aerosol Robotic Network (AERONET);
(6) MetOne beta attenuation monitors operated by USDOS and the Eastern Africa GEOHealth Hub, which are used to calibrate the PurpleAir sensors in order to remove systematic biases.

The Weather Research and Forecasting coupled with Chemistry (WRF-Chem) chemical transport model (CTM) will be used to generate model-based total and speciated PM estimates. GRMs trained by the surface monitors will be used to correct the model outputs for biases, and the results will be merged with the satellite-based PM fields to generate gap-filled maps. Dynamic meteorological data generated by the CTM and static maps of various geographical attributes will be used as auxiliary predictor variables.

4.3 Generated Data Products

A set of standard data products will be generated routinely during the MAIA mission. These include 24-h averaged mass concentrations of total PM_{10}, total $PM_{2.5}$, and $PM_{2.5}$ species (sulfates, nitrates, organic carbon, elemental carbon, and dust) on a 1-km spatial grid for days and locations coincident with satellite instrument observations, as well as spatially and temporally gap-filled versions of these products. All MAIA products will be available free of charge from NASA's Atmospheric Science Data Center (ASDC) (Fig. 2).

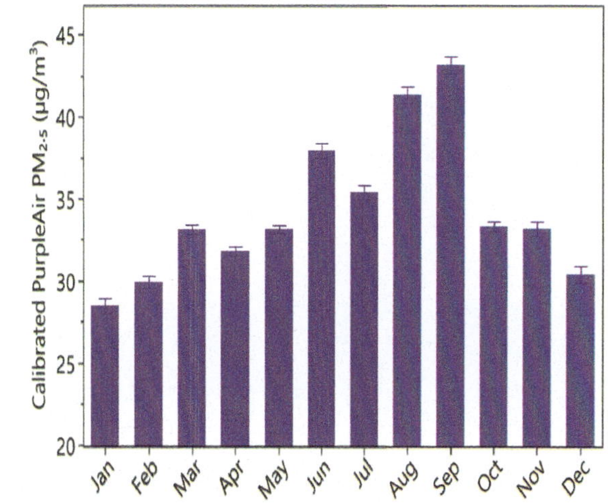

Fig. 2 Monthly $PM_{2.5}$ measurement in Addis Ababa with PurpleAir

4.4 Epidemiological Studies

The MAIA Science Team includes a cadre of epidemiologists and public health experts who will use MAIA products for associating health outcomes and PM exposure on different time scales: acute (short-term), sub-chronic (primarily birth outcomes and pregnancy complications), and chronic (long-term). To protect the privacy of the human subjects whose birth, death, and hospitalization records will be used in MAIA epidemiological studies, none of the products archived at the NASA ASDC will contain health data.

While epidemiological studies assess risk estimates in a population, health impact assessments use this information in combination with other data to attribute outcomes, such as number of asthma cases, to air pollution levels and to explore different policy scenarios. MAIA data will be used to quantify health impact of air pollution in Addis Ababa and Adama using methodologies such as those provided by the WHO AirQ + software tool (Web-7). Findings from the MAIA investigation will be published in the open literature.

5 Conclusion

Satellite data, in conjunction with surface monitors and chemical transport models, provide an integrated approach to assessing exposure to outdoor airborne PM. By helping to characterize the spatial and temporal distribution of PM along with the types of particles that are most harmful to human health, information derived from MAIA investigation can contribute to progress in attaining several Sustainable Development Goals in Ethiopia, including specific societal targets such as reducing the number of deaths and illnesses from air pollution and enhancing international cooperation to facilitate access to clean energy.

References

Diner, D. J., Boland, S. W., Brauer, M., et al. (2018). Advances in multiangle satellite remote sensing of speciated airborne particulate matter and association with adverse health effects: From MISR to MAIA. *Journal of Applied Remote Sensing, 12*, 042603.

Fisher, S., Bellinger, D. C., Cropper, M. L., et al. (2021). Air pollution and development in Africa: Impacts on health, the economy, and human capital. *Lancet Planetary Health, 5*, e681–e688.

GBD 2019 Diseases and Injuries Collaborators. (2020). Global burden of 369 diseases and injuries in 204 countries and territories, 1990–2019: A systematic analysis for the Global Burden of Disease Study 2019. *Lancet, 396*, 1204–1222.

Kumie, A., Worku, A., Tazu, Z., et al. (2021). Fine particulate pollution concentration in Addis Ababa exceeds the WHO guideline value. *Environmental Epidemiology, 8*, e155.

Sanbata, H., Asfaw, A., & Kumie, A. (2014). Indoor air pollution in slum neighbourhoods of Addis Ababa. *Atmospheric Environment, 89*, 230–234.

Tefera, W., Asfaw, A., Gilliland, F., et al. (2016). Indoor and outdoor air pollution-related health problem in Ethiopia: Review of related literature. *Ethiopian Journal of Health Development, 30*, 5–16.

Web-1. https://www.who.int/news/item/04-04-2022-billions-of-people-still-breathe-unhealthy-air-new-who-data.

Web-2. https://www.who.int/news-room/fact-sheets/detail/ambient-(outdoor)-air-quality-and-health.

Web-3. https://www.worldbank.org/en/news/press-release/2021/06/07/report-universal-access-to-sustainable-energy-will-remain-elusive-without-addressing-inequalities.

Web-4. https://sdgs.un.org/goals.

Web-5. https://data.worldbank.org/country/ET.

Web-6. https://et.usembassy.gov/the-u-s-embassy-celebrates-earth-day-with-the-ethiopian-wildlife-and-natural-history-society-installs-air-quality-monitors-in-addis-ababa/.

Web-7. https://www.who.int/europe/tools-and-toolkits/airq---software-tool-for-health-risk-assessment-of-air-pollution.

Aerosol Variability Over Nigeria with a Focus on Changes After COVID-19 Episode

M. S. Shyam Sunder, Dola Tharun, Rajesh Kumar Sahu and Bhishma Tyagi

Abstract

Air pollution threatens human health and the environment locally and globally. Particulate matter ($PM_{2.5}$) monitoring is a crucial component of air quality monitoring, particularly in urbanised and densely populated areas. The COVID-19 pandemic made it very evident that, in such a rare circumstance, the areas of extreme pollution are attracting a higher disease load. The need to check $PM_{2.5}$ concentrations is more significant due to decadal volatility, seasonal variation, or even for the comeback of pollution levels after the pandemic lockdown. The present study focuses on Nigeria to understand the variability and trends of aerosols over the country. Aerosol optical depth (AOD) obtained from MODIS Terra has been analysed spatially for the periods 2010–2019 and 2021–2022 to understand the mean variations and to perform the non-parametric Mann–Kendall trend test. The study also evaluates the in-situ observations of $PM_{2.5}$ concentrations at two of the most populated cities of Nigeria, Abuja and Lagos for 2021. $PM_{2.5}$ concentration level extremity checks were analysed with the value of daily threshold suggested by the Government of Nigeria and World Health Organisation (WHO). The results reveal that the trend of AOD in the area over the past few years has significantly increased, and the daily time series variations are observed more during the dry season. The study suggests that there is a need to reform the existing policies to avoid highly polluted zones in Nigeria.

M. S. Shyam Sunder · D. Tharun · R. Kumar Sahu · B. Tyagi (✉)
Department of Earth and Atmospheric Sciences, National Institute of Technology Rourkela, Rourkela, Odisha 769008, India
e-mail: tyagib@nitrkl.ac.in

Keywords

Aerosols · Mann–Kendall · MODIS · $PM_{2.5}$ · Nigeria air quality

1 Introduction

Pollution is one of the most significant hazards to human health (Nwani, 2022). Exposure to air pollution over time has grown dangerous for human mortality, especially for babies, as it increases lung-related illnesses and may worsen pandemic conditions in the extremes (Etchie et al., 2021; Sahu et al., 2022; Tyagi et al., 2022). In light of prior studies on air quality, it is essential to consistently ensure that the air we breathe is not leading the risk of burden of disease or pre-mature mortality (Kamyotra et al., 2011; CPCB, 2023). It is crucial to accurately evaluate particulate matter (PM), one of the main elements of air pollution, because of significant correlations between environmental emissions and industrial/economic growth (Sulaymon et al., 2020; Jin et al., 2017). Various elements, such as dust, smoke, organic matter, gas pollutants, and air particles, make up the primary or secondary PM (Zhou & Wang, 2018). According to Ogunjobi and Kim, the diameter of PM can range from a few nanometres (nm) to tens of micrometres (m), which is harmful to living creatures (Krewski et al., 2003; Taus et al., 2008).

The average individual inhales daily doses of fine particulate matter, also known as $PM_{2.5}$, which has a width of 2.5 m, for which the daily threshold by WHO is 25 $\mu g/m^3$ (Gogikar & Tyagi, 2016; Opara et al., 2021). Smaller with a wide surface area, $PM_{2.5}$ also reduces visibility as concentration rates rise (Cao & Han, 2015). It damages the environment, the human body, and biological ecosystems (Fang et al., 2016). The research on $PM_{2.5}$ includes various relative perceptions, e.g., the effect of meteorological

K. W. Fomba et al. (eds.), *Advances in Air Quality Research in Africa*,
Advances in Science, Technology & Innovation, https://doi.org/10.1007/978-3-031-53525-3_21

conditions on $PM_{2.5}$ concentrations (Su et al., 2016). $PM_{2.5}$ concentration time series changes over local localities (Zhao et al., 2014) analysis of spatial distribution (Zhang et al., 2013) and impact of $PM_{2.5}$ on health (Barudgar et al., 2022; Tyagi et al., 2022). The interaction of chemicals in the air, caused by human anthropogenic fires, fuel combustion, wood stoves, soot, and forest fires, is the leading cause of $PM_{2.5}$ generation (Shaibu & Weli, 2017; Tariq et al., 2022). Variations in the height of the boundary layer in the atmosphere impacted regional $PM_{2.5}$ concentrations, which governed synoptic airflow and vertical mixing of pollutants (Shaibu & Weli, 2017). The United States Environmental Protection Agency (USEPA), one of the leading environmental agencies, conducts routine air quality monitoring and offers recommendations for controlling air quality. These initiatives have significantly decreased $PM_{2.5}$ levels in Western Europe and North America (Balogun & Odjugo, 2022; Mackie et al., 2016; Ragothaman & Anderson, 2017).

The African population is open to greater exposure to increasing pollution, which results in severe health issues (Pearson et al., 2010; Pope & Dockery, 2006; Shindell et al., 2022). Air pollution understanding is crucial in fast-developing African regions, such as Nigeria, which has risen in its industrial growth tremendously since the industrial revolution (Shaibu & Weli, 2017). Nigeria has the most significant population in Africa and is the third-largest industrial sector, and the economy is expanding quickly. Food, beverages, tobacco, plastics, tree products, papers, waste recycling, metals, motors, cars, electrical equipment, and electronics are all produced in Lagos to a greater than 70% extent (Nwagwu & Oni, 2015; Pure, 2022). However, due to limitations in the temporal and spatial frequency of data collection, there are certain limitations in understanding the $PM_{2.5}$ consequences in the African region (Petkova et al., 2013). Another popular variable for measuring the aerosol impact on any location is Aerosol Optical Depth (AOD), which is a satellite-based value. Due to wide coverage of satellite observations, AOD is a useful parameter, obtained from various onboard instruments on different satellites, and can give a global spatial map of the variable.

The present study uses satellite measurements of aerosol optical depth (AOD) to discuss the spatial distribution of aerosols over Nigeria from 2010 onwards. The current study also analyses in-situ recorded $PM_{2.5}$ levels post-COVID-19 lockdowns over Lagos and Abuja cities of Nigeria. The primary focus is on understanding the (i) Aerosol Optical Depth (AOD) variation for a decade (2010–2019) and identifying the significant air pollution concentration hotspots over Nigeria, (ii) Daily and seasonal variation of the $PM_{2.5}$ for Abuja and Lagos locations.

2 Study Area, Data Description and Methodology

Figure 1 depicts Nigeria (the Gaint of Africa), which is in the West of Africa. It features plains, marshes, mountains, and deserts, as well as one of the world's most unique river systems and the third-largest delta, covered in plains and savannas. According to United Nations estimations, Nigeria has a population of 213,401,323 and a density of 167.5 persons per square kilometre (Peng-In et al., 2022). Nigeria has two distinct seasons, the summer and the Harmattan season, which last from April to October and November to March, respectively (Balarabe et al., 2015).

This study uses the Moderate Resolution Imaging Spector-radiometer (MODIS), with the constellation of Aqua and Terra (Peng-In et al., 2022); day and night sensors data products, with a nominal spatial resolution of 10 km^2, to analyse the spatial distribution of trends of Aerosol Optical Depth (AOD) with 550 nm wavelength. This study employs Mann–Kendall (MK) trend, a non-parametric statistical analysis for the time series data resulting in both positive and negative trend values to compute the AOD trends at the confidence level of 95% (Hari et al., 2022; Sahu & Tyagi, 2022; Vissa et al., 2019).

We used Abuja and Lagos, Nigeria's most populous and developed cities, as our analysis's points of reference. Daily $PM_{2.5}$ in-situ observations were adapted from the U.S. Embassy (AirNow) for this study over Abuja and Lagos. To analyse the daily and seasonal fluctuations of $PM_{2.5}$ over the stations, the Government of Nigeria and WHO (2021) set threshold values for daily $PM_{2.5}$ concentrations of 35 and 25 $\mu g/m^3$. We have used daily AOD values over Nigeria for 2010–2022, whereas the in-situ observed values of $PM_{2.5}$, acquired from U.S. Embassy are for January 2021–December 2021.

3 Results and Discussion

3.1 Aerosol Variability Over Nigeria

The variability of aerosols has been a prime topic of research in air quality over Nigeria. Climate, urbanisation, population, economic growth, and industrial development influence air pollution and emission levels in Africa (Gahungu et al., 2022). From the existing studies of Nigeria, it is reported that the wet seasons (summer) have better, and cleaner air quality as compared to the dry (harmattan) seasons (Kanee et al., 2020). Various studies using in-situ, satellite and model simulations assessed the aerosol

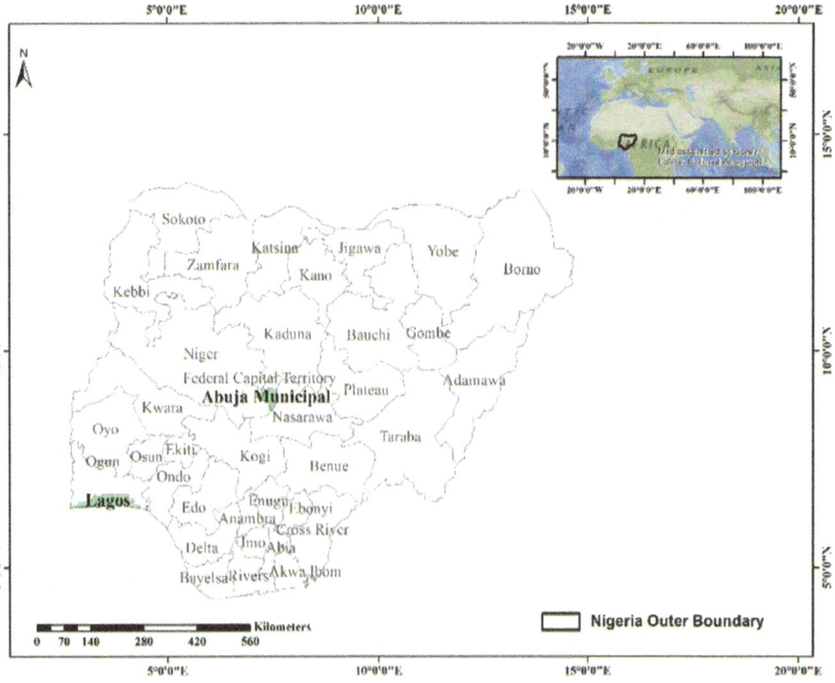

Fig. 1 Study area location of Nigeria with Abuja and Lagos location is highlighted in bold letters with green-coloured fills with the state boundary names

concentrations/variability over different sites of Nigeria, of which some are summarised in Table 1. The main inference of all the mentioned studies is having higher values than thresholds of aerosols over Nigeria for the past many years during different seasons.

To further see the variability over a spatial scale, we have used MODIS Aqua-Terra combined AOD values for whole country for 2010–2022. Figure 2 represents the average spatial variation of AOD for the years 2021–2022 and 2010–2019, along with difference of the previous and later. One can make a subtle observation that the average AOD values are higher for 2021–2022 compared to average plot of 2010–2019. The variability demonstrates that the coastline accumulated greater AOD values in the southern and northern Nigeria.

This vegetation loss is attributed to anthropogenic activities (human influence) and climatic changes. When land degradation patterns were identified, notably in the Sub-Saharan area, it was discovered that there had been a large-scale depletion due to an unprecedented change in landuse (Globalforestwatch, 2023). Sahelian drought and climate change, especially those triggered by human activity, are well-known in this region (Adeaga et al., 2022).

The central region of Nigeria has a mid-altitude Eco-climate zone with average AOD ranges of less than 0.3 near the plateau state. AOD patches of lower value are also present in areas that are close to the ocean. When comparing the extremes of both north and south of Nigeria, the northwestern zones received an AOD that ranged between 0.35 and 0.7 (near Jigawa state), revelling a radial pattern, while the south Nigeria received an AOD between 0.45 and 0.7 for 2021 to 2022 and between 0.4 and 0.65 for 2010 to 2019 (Fig. 2).

The third column of Fig. 2 illustrates the spatial difference in AOD between 2021–2022 and 2010–2019, which aids in understanding the areas where the AOD extends higher and lower. The difference in AOD is positive, indicating a relative rise in the AOD levels, particularly over the Southern Guinea Savannas zone in the last 2 years, which are associated with greater urban growth (Adeaga et al., 2022). The borders of the NW and the south along the coastal boundaries had positive and negative values in distributions. In contrast, the middle NE and NW regions had positive variation and the zonal regions near southern coast of Nigeria (Lagos, south part of Edo, Delta, Bayelsa Rivers, Imo and Abia Akwa Ibom) observed to have positively high variations up to 0.2, according to the difference in the spatial map.

3.2 Spatial Trend Variability of AOD Over Nigeria

Figure 3 depicts the MK trend performed over Nigeria with 95% significance (shown in black dots) using the MODIS AOD product for 2010–2019. For the decadal trend of AOD (for 2010–2019), Nigeria shows decreasing trend values of AOD over southern region (ranging from ~−0.2 to −0.7), and over the state of Borno (ranging from −0.04 to −0.08), whereas all the other states of Nigeria are showing either no trend or a slightly positive trend of AOD. We found that the NE and NW regions of the northern states, which are prone to desertification, had a high and low positive trend of AOD.

Table 1 Mean concentrations of PM over Nigeria observed during different studies

Serial number	Authors year and study area	Data description	Main findings
1	Ayodele et al. (2021): Lagos	$PM_{1.0}$, $PM_{2.5}$, and PM_{10} data with weekly measurement of 9 AM to 5 PM from February–July 2019	The Aerosol optical Thickness (AOT) values increased from 2002 to 2020, which might be attributed to the metropolis's rising urbanisation
2	Oluyemi and Asubiojo, (2001): Lagos	Air particulate samples collected at Ikeja in Lagos, Yaba premises of the Federal Ministry of Health Central Laboratory and Eti-Osa in Ikoyi for 1989–1991	Four sources stand out as significant contributors to the air particulate loading in the Lagos metropolitan, according to the factor analysis findings of the three locations in Lagos. They are incineration, soil, marine, and vehicle traffic
3	Njoku et al. (2016): Lagos	CO, CO_2, O_2, SO_2, NO_X, and NO and Suspended Particulate Matter (SPM) for one hour of exposure time with twenty-one random samplings	The study's results indicated that levels of NO, NO_2, CO, and SPM in all sample regions were relatively high and beyond regulation limits, but there was no significant change in any of these parameters except for SPM (at all sampling stations)
4	Obisesan, 2021: Ile-Ife	Aerosol optical depth (AOD) for time period 12:30–01:30 pm from November 2017 to March 2018	The study location experiences a polluted atmosphere during the Harmattan season
5	Baumbach et al. (1995): Lagos	NOx, O_3, SO_2, CO, CO_2, wind speed, direction, temperature, relative humidity, rain duration, and solar density for the time period 1993–1994	During the dry season, the Lagos district region noted large concentrations of Sahara dust particle matter, high volumes of CO and SO, and other combustion by-products. However, high-traffic regions tend to have the most severe issues
6	Obioh et al. (2013): Lagos	$PM_{2.5}$ and PM_{10} samplings collected at 6:00 to 11:00, 12:00 to 15:00 and 16:00 to 21:00 during September and October 2009	The study revealed that all receptor locations where PM_{10} mass loads were measured violated the WHO threshold limit recommendation. Most cities had $PM_{2.5}$ mass loads that exceeded the WHO's suggested threshold limit guideline. $PM_{2.5}/PM_{10}$ ratio was comparable to that of a moderately polluted major city

(continued)

Table 1 (continued)

Serial number	Authors year and study area	Data description	Main findings
7	Owoade et al. (2013): Lagos	Coarse ($PM_{2.5}$) and fine ($PM_{2.5}$-PM_{10}) airborne particulate matter (October 2010, 7:00 am to 7:00 pm)	Except for Victoria Island, sample locations showed a substantial variation in particle loads on weekdays and weekends
8	Meseke et al. (2022): Kano, Abuja	Particulate matter ($PM_{1.0}$, $PM_{2.5}$, and $PM_{10.0}$), Temperature, relative humidity for 1st January–31st December 2021	The results indicate that the monthly average PM_1, $PM_{2.5}$, and PM_{10} concentrations at both sites are higher than the WHO 24-h standard limit during the dry season and below the WHO 24-h guideline during rainy season. Both areas are among the most productive zones in Nigeria's northern region, with expanding population
9	Akinfolarin et al. (2017): Harcourt	PM_{10} and $PM_{2.5}$ and Automated GPS for sampling point coordinates (2016)	The dry season concentrations were > 150 µg/m³ and 230 µg/m³. For the rainy season, the three new industrial sites' air quality indexes ranged from "good" to "moderate". In contrast, they ranged from "very unhealthy" to "hazardous" during the dry season
10	Shehu et al. (2019): Abuja	ECMWF data over Abuja, Nigeria (2017)	The study shows that the annual increase in the human population had a detrimental effect on aerosol concentrations. The aerosol concentrations are higher in the dry season than in the rainy season
11	Sulaymon et al., 2020: Lugbe and Abuja	At two locations (NP and WE) in the Lugbe region, ambient $PM_{2.5}$ was obtained (2016)	The 120-h backward trajectories indicate that Abuja's northeast dominated in January while the southwest dominated for the following three weeks

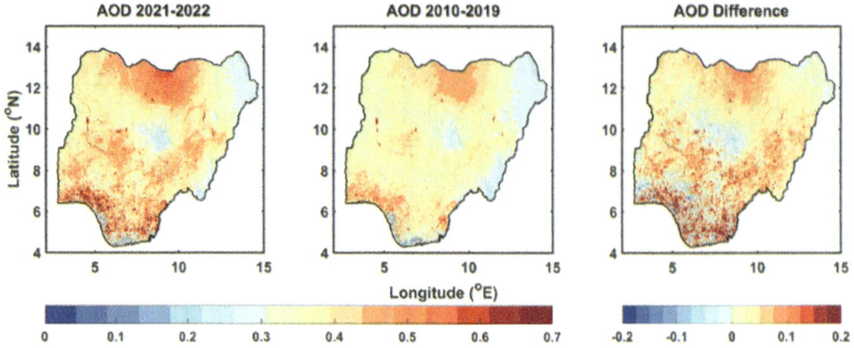

Fig. 2 Spatial variations of AOD average for the year 2021–2022 (First Column), AOD average for the year 2010–2019 (Second Column) and the spatial difference with both spatial averages (Third Column)

Compared to NW, the NE region values range between 0.04 and −0.14. The peak value of the 2010–2019 trend was in the NW of Nigeria, and the lowest trend values dropped as low nearby −0.03 and the highest was 0.08. Overall, the lowest and no trend scenario was seen with the MK trend in Nigeria's NE and NW boundaries. However, it will be sufficient to mention that the southern Nigeria is showing a reduction in AOD values whereas the rest of the Nigeria is experiencing a rise (positive) in trend values of AOD over the period, though different regions have different intensities of increasing trend values. Here we have not included during COVID or post-COVID periods in the trend analysis due to drastic changes of AOD loading worldwide with reduced and enhanced values. Hence, the trend analysis is limited for one decade (2010–2019) only over Nigeria.

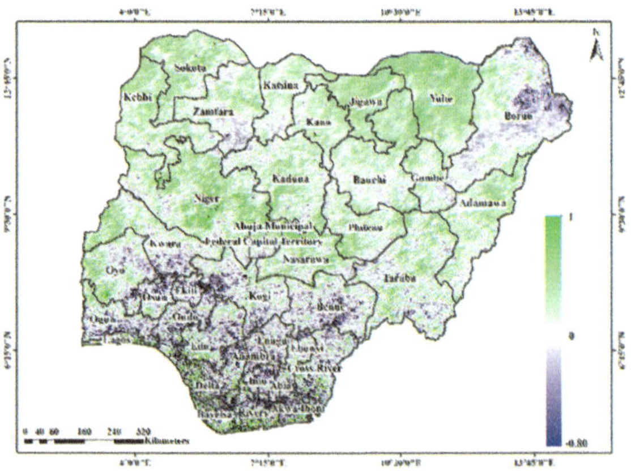

Fig. 3 Spatial trend map using Mann–Kendall approach for 2010–2019 for Nigeria for annual trends in AOD. The 95% significance regions are shown with black dots

3.3 Seasonal Variability of $PM_{2.5}$ Using In-Situ Observations

The in situ observations of $PM_{2.5}$ have been collected over Abuja and Lagos for 2021. The daily concentrations of fine particulate matter ($PM_{2.5}$) for Abuja and Lagos regions fluctuate with season (Fig. 4). Figure 4a shows the $PM_{2.5}$ variability during summer season (April–October). It is observed that for initial few days, Abuja has higher concentrations of $PM_{2.5}$, which will eventually reach to the threshold fixed by Govt. of Nigeria for the season. However, for the Lagos, there are many days crossing the safe threshold limit during the summer season.

$PM_{2.5}$ levels exceed the standards set by the Nigerian government for April, June, July and October for Abuja. Due to the wet deposition of suspended particulate matter and the impact of aerosol hygroscopic development, such a reduction or maintenance in the $PM_{2.5}$ concentration level is anticipated. Nevertheless, it exceeded both standards in Lagos, with the concentrations exceeding 55 μg/m^3 in June. The low humidity, high temperature and less rainfall of the Harmattan season (November to March) create a desert-like environment that encourages the formation of dust and sandstorms (Weinstein et al., 2010) (Fig. 4b).

The $PM_{2.5}$ concentrations, especially in the second half of January and February, from July to October, were above WHO and Nigeria guidelines, exhibiting a significant increase in $PM_{2.5}$ concentrations in July and a subsequent decline in October. The ranges of $PM_{2.5}$ concentration variation were shown to be substantially more remarkable during the dry and dusty seasons, but they were much narrower during the rainy season. The northeasterly trade winds that blow from the Sharan desert over West Africa and into the Gulf of Guinea were accompanied by bigger cloud formations that included significant dust. For the complete year 2021 (Fig. 2c), there was a substantial rise in the rate of $PM_{2.5}$ within February and April for Abuja, which is the initial phase of the summer season.

However, the $PM_{2.5}$ over Lagos followed the sequential rise and stalled with an exponential fall in January and February. In February, Lagos got the peak $PM_{2.5}$ value of ~330 μg/m^3, which is more than Abuja's peak value.

Table 2 shows the seasonal variation of mean and standard deviations over Abuja and Lagos. According to Table 2, both Abuja and Lagos had greater dispersion from the mean during the harmattan season. The average $PM_{2.5}$ concentrations over Lagos (Abuja) during the Harmattan

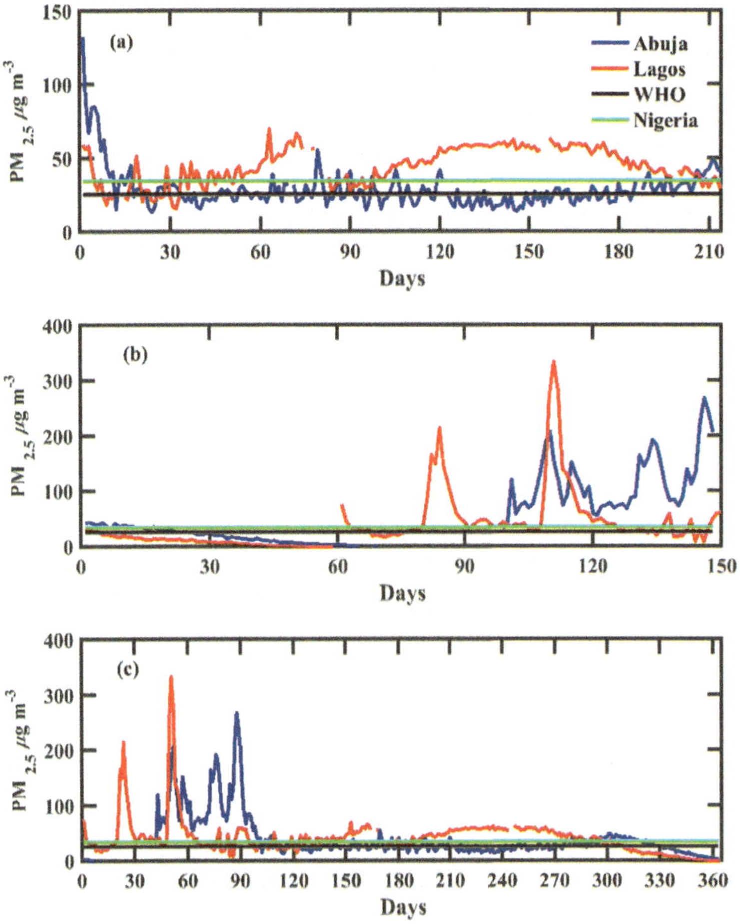

Fig. 4 Seasonal and annual variation of $PM_{2.5}$ concentration over Abuja and Lagos (Nigeria) **a** variation of $PM_{2.5}$ in summer season (April–October), **b** Variation of $PM_{2.5}$ concentration during Harmattan season (November–March), **c** Variation of $PM_{2.5}$ in 2021. The *x*-axis represents days in each subplot

Table 2 $PM_{2.5}$ seasonal mean and standard deviation for Abuja and Lagos location

Location	Season	Mean (μg/m^3)	Standard deviation (μg/m^3)
Abuja	Summer	28.97	13.71
	Harmattan	60.38	60.26
Lagos	Summer	45	12.15
	Harmattan	38.33	50.99

season is 50.99 μg/m3 (60.38), and the standard deviation is 38.33 μg/m3 (60.26), respectively. It is evident that Lagos experiences greater volatility than Abuja. The impact of higher $PM_{2.5}$ values affects the region severely, with more premature deaths reported by the World Bank in Abuja and Lagos (Gahungu et al., 2022).

4 Conclusion

The variability of aerosol shows that Nigeria is fast growing and the AOD values are higher over most of the country. The study utilised AOD values for the last decade and the surface observation for 1-year period. The findings can be summarised as follows:

- The spatial average of AOD for the decade (2010–2019) shows higher AOD concentration over the northern region near the Desert with radial pattern, having more positive AOD density near northern boundaries of Jigawa state in the trend. The spatio-temporal trend of AOD indicates a decreasing trend over the southern Nigeria and over Borno, whereas the rest of country didn't indicate any significant trend values.

- The AOD values are showing a surge over most of the states of Nigeria after COVID-19 lockdown (for year 2021–2022). The change can be understood with fast growing economy and transportation in the recent years.
- Seasonal analysis reveals that during the summer season, the concentration of $PM_{2.5}$ was substantially high in April, exceeding both WHO and Nigeria's daily $PM_{2.5}$ threshold criteria. On the other hand, during the Harmattan season, the concentration of $PM_{2.5}$ was observed below the threshold values of WHO and Nigeria's daily $PM_{2.5}$ concentrations from November to January, exceeding the threshold values after February.

Acknowledgements The authors would like to acknowledge the National Institute of Technology Rourkela for providing lab facilities. The authors express sincere thanks to scientists working with AirNow (https://www.airnow.gov/) and Giovanni (https://giovanni.gsfc.nasa.gov/giovanni/) for providing the datasets needed to do the present work.

References

Adeaga, O., Lawal, O., Adedeji, O., & Akinbaloye, O. (2022). Assessment of vegetation cover dynamics in the agro-ecological Zones of Nigeria. *Geogr. Phys. Geogr., 22*, 19–32. https://doi.org/10.12775/bgeo-2022-0002

Airnow, Retrieved January 14, 2023, from http://www.airnow.gov.

Akinfolarin, O., et al. (2017). Assessment of particulate matter-based air quality index in Port Harcourt Nigeria. *Journal of Environmental Analytical Chemistry, 4*, 1–4. https://doi.org/10.4172/2380-2391.1000224

Ayodele, E., et al. (2021). Using remote sensing to monitor Aerosol Optical Thickness (AOT) and its relationship with land cover in Lagos Metropolis. *Nigeria. Intercontinental Geoinformation, 3*, 117–120.

Balarabe, M., et al. (2015). Seasonal variations of aerosol optical properties and identification of different aerosol types based on AERONET data over sub-Sahara West-Africa. *Atmospheric and Climate Sciences., 6*, 13–28. https://doi.org/10.4236/acs.2016.61002

Balogun, V., & Odjugo, P. (2022). Spatial Analyses of Air Pollutants Concentration around the Warri Refining and Petrochemical Company (WRPC), Delta State. *Nigeria. Ghana Journal of Geography., 14*, 50–81. https://doi.org/10.4314/gjg.v14i2.3

Barudgar, A. et al. (2022). Variability of fine particulate matter ($PM_{2.5}$) and its Association with health and vehicular emissions over an urban tropical coastal station Mumbai, India. *Thalassas: International Journal of Marine Science,* 1–4. https://doi.org/10.1007/s41208-022-00442-4, https://doi.org/10.1007/s41208-022-00442-4.

Baumbach, G., et al. (1995). Air pollution in a large tropical city with a high traffic density—results of measurements in Lagos Nigeria. *Science of the Total Environment, 169*, 25–31. https://doi.org/10.1016/0048-9697(95)04629-F

Cao, C., & Han, L. (2015). The assessment on the social health costs caused by fog and haze. Presented at the June 14, Stat Res.

CPCB Webpage. Retrieved January 14, 2023 from https://cpcb.nic.in/.

Etchie, T., et al. (2021). Season, not lockdown, improved air quality during COVID-19 State of Emergency in Nigeria. *Science of the Total Environment, 768*, 145187. https://doi.org/10.1016/j.scitotenv.2021.145187

Fang, C., Wang, Z., & Xu, G. (2016). Spatial-temporal characteristics of PM2.5 in China: A city-level perspective analysis. *Journal of Geographical Sciences, 26*, 1519–32. https://doi.org/10.1007/s11442-016-1341-9.

Gahungu, P., Kubwimana, J., Muhimpundu, L., & Ndamuzi, E. (2022). Modelling spatio-temporal trends of air pollution in Africa. https://doi.org/10.48550/arXiv.2208.12719.

Gogikar, P., & Tyagi, B. (2016). Assessment of particulate matter variation during 2011–2015 over a tropical station Agra India. *Atmospheric Environment, 147*, 11–21. https://doi.org/10.1016/j.atmosenv.2016.09.063

Hari, M., Sahu, R., Sunder, M., Tyagi, & B. (2022). Then and Now: COVID-19 Pandemic Lockdown Misfire Atmospheric Methane over India. *Aerosol and Air Quality Research*, 210354. https://doi.org/10.4209/aaqr.210354.

Jin, Q., Fang, X., & Wen, B. (2017). Shan: A Spatio-temporal variations of $PM_{2.5}$ emission in China from 2005 to 2014. *Chemosphere, 183*, 429–36. https://doi.org/10.1016/j.chemosphere.2017.05.133

Kamyotra, J., Saha, D., Tyagi, S., Sen, A., Srivastava, R., Pathak, A. (2011). *Guidelines for the measurement of ambient air pollutants.* Central Pollution Control Board-Ministry of Environment & Forests, Govt. of India.

Kanee, R., Adeyemi, A., Edokpa, D., & Ede, P. (2020). Particulate matter-based air quality index estimate for Abuja, Nigeria: Implications for health. *Journal of Geoscience and Environment, 8*, 313–321. https://doi.org/10.4236/gep.2020.85019

Krewski, D., Burnett, R., Goldberg, M., Hoover, B., Siemiatycki, J., Jerrett, M., Abrahamowicz, M., & White, W. (2003). Overview of the reanalysis of the Harvard six cities study and American Cancer Society study of particulate air pollution and mortality. *Journal of Toxicology and Environmental Health, Part A, 66*, 1507–1552. https://doi.org/10.1080/15287390306424

Mackie, A., Palmer, P., Barlow, J., Finch, D., Novelli, P., & Jaeglé, L. (2016). Reduced Arctic air pollution due to decreasing European and North American emissions. *Journal of Geophysical Research: Atmospheres, 121*, 8692–8700. https://doi.org/10.1002/2016JD024923

Meseke, N., Akpootu, D., Falaiye, O., Targema, T.: Comparative Assessment of Particulate Matter Using Low Cost Sensor: A Case Study of Abuja And Kano, Nigeria. Fudma J. Sci. 6, 203–11 (2022). https://doi.org/10.33003/fjs-2022-0604-1066.

Njoku, K., Rumide, T., Akinola, M., Adesuyi, A., & Jolaoso, A. (2016). Ambient air quality monitoring in metropolitan city of Lagos, Nigeria. *Journal of Applied Sciences and Environmental Management., 20*, 178–185.

Nwagwu, I., Oni, T. (2015). Lagos and its potentials for economic growth. Heinrich Böll Stiftung.

Nwani, S. (2022). Air pollution trajectories and life expectancy in Nigeria. *International Journal of Social Economics, 49*, 1049–1070. https://doi.org/10.1108/IJSE-06-2021-0351

Obioh, I., Ezeh, G., Abiye, O., Alpha, A., Ojo, E., & Ganiyu, A. (2013). Atmospheric particulate matter in Nigerian megacities. *Toxicological and Environmental Chemistry, 95*, 379–385. https://doi.org/10.1080/02772248.2013.790970

Obisesan, O. (2021). Aerosol optical depths during two harmattan seasons in Ile-Ife Nigeria. *Atmosphere., 11*, 150–161. https://doi.org/10.9734/IJECC/2021/v11i630431

Ogunjobi, K., & Kim, Y. (2008). Aerosol characteristics and surface radiative forcing components during a dust outbreak in Gwangju, Republic of Korea Republic of Korea. *Environmental Monitoring and Assessment, 137*, 111–126. https://doi.org/10.1007/s10661-007-9733-z

Oluyemi, E., & Asubiojo, O. (2001). Ambient air particulate matter in Lagos, Nigeria: A study using receptor modeling with x-ray flourescence analysis. *Bulletin of the Chemical Society of Ethiopia, 15*, 97–108.

Opara, A., Akaolisa, C., Akakuru, C., Nkwoada, A., Ibe, F., Verla, A., Chukwuemeka, I. (2021). Particulate matter exposure and noncancerous inhalation health risk assessment of major dumpsites of Owerri metropolis, Nigeria. *Environmental Analysis Health and Toxicology, 36*. https://doi.org/10.5620/eaht.2021025

Owoade, O., Fawole, O., Olise, F., Ogundele, L., Olaniyi, H., Almeida, M., Ho, M., & Hopke, P. (2013). Characterization and source identification of airborne particulate loadings at receptor site-classes of Lagos Mega-City Nigeria. *Journal of the Air & Waste Management Association, 63*, 1026–1035. https://doi.org/10.1080/10962247.2013.793627

Pearson, J., Bachireddy, C., Shyamprasad, S., Goldfine, A., & Brownstein, J. (2010). Association between fine particulate matter and diabetes prevalence in the US. *Diabetes Care, 33*, 2196–2201. https://doi.org/10.2337/dc10-0698

Peng-In, B., Sanitluea, P., Monjatturat, P., Boonkerd, P., & Phosri, A. (2022). Estimating ground-level $PM_{2.5}$ over Bangkok Metropolitan Region in Thailand using aerosol optical depth retrieved by MODIS. *Air Quality, Atmosphere and Health, 15*, 2091–102. https://doi.org/10.1007/s11869-022-01238-4

Petkova, E., Jack, D., Volavka-Close, N., & Kinney, P. (2013). Particulate matter pollution in African cities. *Air Quality, Atmosphere and Health, 6*, 603–614. https://doi.org/10.1007/s11869-013-0199-6

Pope, C., III., & Dockery, D. (2006). Health effects of fine particulate air pollution: Lines that connect. *Journal of the Air and Waste Management Association, 56*, 709–742. https://doi.org/10.1080/10473289.2006.10464485

Pure, N. (2022). Spatial and temporal variation of particulate matter around Lagos environs. *Nigerian Journal of Pure and Applied Science, 35*, 4310–4321. https://doi.org/10.48198/NJPAS/22.A17

Ragothaman, A., & Anderson, W. (2017). Air quality impacts of petroleum refining and petrochemical industries. *Environments, 4*, 66. https://doi.org/10.3390/environments4030066

Sahu, R., Tyagi, B., Vissa, N., Mohapatra, M. (2022). Pre-monsoon thunderstorm season climatology of convective available potential energy (CAPE) and convective inhibition (CIN) over eastern India. *73*, 565–86. https://doi.org/10.54302/mausam.v73i3.1247.

Sahu, R., & Tyagi, B. (2022). Spatial variation of thermodynamic indices over north-east India during pre-monsoon thunderstorm season. *Journal of Atmospheric and Solar - Terrestrial Physics, 232*, 105868. https://doi.org/10.1016/j.jastp.2022.105868

Shehu, M., Umaru, I., & Tukura, B. (2019). Health impact analysis of some outdoor atmospheric aerosols ($Pm_{2.5}$, SO_2, CO & CH_4) in FCT Abuja and environs Nigeria. *American Journal of Public Health, 7*, 1–8.

Shaibu, V., & Weli, V. (2017). Relationship between $PM_{2.5}$ and climate variability in Niger delta Nigeria. *American Journal of Environmental Protection, 5*, 20–4. https://doi.org/10.12691/env-5-1-4

Shindell, D., Faluvegi, G., Parsons, L., Nagamoto, E., & Chang, J. (2022). Premature deaths in Africa due to particulate matter under high and low warming scenarios. *GeoHealth*. https://doi.org/10.1029/2022GH000601

Su, X., Gough, W., & Shen, Q. (2016). Correlation of $PM_{2.5}$ and meteorological variables in Ontario cities: statistical downscaling method coupled with artificial neural network. *WIT Transactions on Ecology and the Environment, 207*, 215–26.

Sulaymon, I., Mei, X., Yang, S., Chen, S., Zhang, Y., Hopke, P., Schauer, J., & Zhang, Y. (2020). $PM_{2.5}$ in Abuja, Nigeria: Chemical characterization, source apportionment, temporal variations, transport pathways and the health risks assessment. *Atmospheric Research, 237*, 104833. https://doi.org/10.1016/j.atmosres.2019.104833

Tariq, S., Nawaz, H., Mehmood, U., Babar, Z. (2022). Remote sensing of aerosols due to biomass burning over Kanpur, Sao-Paulo, Ilorin and Canberra. *Journal of Atmospheric Chemistry*, 1–52. https://doi.org/10.1007/s10874-022-09444-1.

Taus, N., Tarulescu, S., Idomir, M., & Taus, R. (2008). Respiratory exposure to air pollutants. *Journal of Environmental Protection and Ecology, 9*, 15–25.

Tyagi, B., Vissa, N., & Ghude, S. (2022). Evolution of pollution levels from COVID-19 lockdown to post-lockdown over India. *Toxic., 10*, 653. https://doi.org/10.3390/toxics10110653

Vissa, N., Anandh, P., Behera, M., & Mishra, S. (2019). ENSO-induced groundwater changes in India derived from GRACE and GLDAS. *Journal of Earth System Science, 128*, 1–9. https://doi.org/10.1007/s12040-019-1148-z

Globalforestwatch: Vizzuality, Forest monitoring, land use & deforestation trends. Retrieved January 7 2023, from https://www.globalforestwatch.org

Weinstein, J., Hedges, S., & Kimbrough, S. (2010). Characterization and aerosol mass balance of $PM_{2.5}$ and PM_{10} collected in Conakry, Guinea during the 2004 Harmattan period. *Chemosphere, 78*, 980–8. https://doi.org/10.1016/j.chemosphere.2009.12.022

WHO. (2021). World Health Organization global air quality guidelines: particulate matter ($PM_{2.5}$ and PM_{10}), ozone, nitrogen dioxide, sulfur dioxide and carbon monoxide. Genève, Switzerland: World Health Organization.

Zhang, R., Jing, J., Tao, J., Hsu, S., Wang, G., Cao, J., Lee, C., Zhu, L., Chen, Z., Zhao, Y., & Shen, Z. (2013). Chemical characterization and source apportionment of $PM_{2.5}$ in Beijing: Seasonal perspective. *Atmos Chem Phys., 13*, 7053–74. https://doi.org/10.5194/acp-13-7053-2013

Zhao, C., Wang, Y., Wang, Y., Zhang, H., Zhao, B. (2014). Temporal and spatial distribution of $PM_{2.5}$ and PM_{10} pollution status and the correlation of particulate matters and meteorological factors during winter and spring in Beijing. *Huan jing ke xue = Huanjing kexue, 35*, 418–27.

Zhou, C., Li, S., & Wang, S. (2018). Examining the impacts of urban form on air pollution in developing countries: A case study of China's megacities. *International Journal of Environmental Research and Public Health, 15*, 1565. https://doi.org/10.3390/ijerph15081565

Ecohealth Chair in Urban Air Pollution and Its Impact on Non-communicable Respiratory Diseases in West Africa

Nonvignon Marius Kêdoté, Aymeric Joaquin Darboux, Issaka Tiembre, Sandrine Lompo and Mamadou Fall

Abstract

The "Ecohealth Chair on Urban Air Pollution and Non-Communicable Respiratory Diseases in West Africa" is a regional project whose aim is to promote the health and well-being of populations in a healthy environment in four cities (Abidjan, Cotonou, Dakar and Ouagadougou). Through an interdisciplinarity approach, the research components contribute to characterize and measure the health risks related to air pollution in various vulnerable and particularly exposed groups (children, women, motorcycle cab drivers, bus drivers, shopkeepers living along the roads). Indoor and outdoor pollution levels measured in urban West Africa are high with greater concentration at crossroads. The main risk factors are the intensity of urban traffic, the use of biomass, the location of the kitchen inside the dwelling, the low number of windows in the rooms, and the use of products to control disease-carrying insects and improve olfactory sensations. Also, the level of implementation of air quality regulations has been accessed. Important mechanisms and institutions exist to protect air quality but they encounter obstacles related to governance and resources made available. In addition, some legal provisions appear to be outdated and unknown to the actors. In the action component, an intervention called Air-sain has been implemented and allowed, in addition to mass actions, to encourage the use of less polluting energy sources and to influence positively a thousand community and institutional actors. The "Ecohealth Chair" project has, through these different axes, raised local awareness of the problem of air pollution and initiated a change in the practices of communities and institutions.

Keywords

Air quality · Non-communicable diseases · Community intervention · West Africa

1 Introduction

The most economically and health-vulnerable communities are often unaware of the public health problems associated with poor air quality (Odonkor & Mahami, 2020). Thus, the lack of data on air pollution, its health effects, and the influence of lifestyle needed to be addressed by generating air quality monitoring data, health information, and associated socio-economic data (Amegah & Agyei-Mensah, 2017; Katoto et al., 2019). Although in recent years, public policies tend to take more into account the problem of urban pollution, efforts are still needed to ensure the effective implementation of the proposed mitigation and prevention measures. Indeed, significant limitations in terms of the operationality and effectiveness of some of these mitigation measures in West African countries remain

N. M. Kêdoté (✉)
Département Santé Environnement, Institut Régional de Santé Publique Comlan Alfred Quenum (IRSP-CAQ), Université d'Abomey-Calavi, Abomey Calavi, Bénin
e-mail: kedmar@yahoo.fr

N. M. Kêdoté · A. J. Darboux
Communauté de Pratique Écosanté et Santé Humaine en Afrique de l'Ouest et du Centre (COPES-AOC), Cotonou, Bénin

I. Tiembre
UFR SMA, Département Santé Publique, Chaire ECOSANTE, West Africa GEOHealth Network, Université Félix Houphouët Boigny, Abidjan, Côte d'Ivoire

S. Lompo
Laboratoire de Santé Publique, Chaire ECOSANTE, West Africa GEOHealth Network, Université Pr. Joseph Ki-Zerbo, Ouagadougou, Burkina Faso

M. Fall
Laboratoire de Toxicologie et Hydrologie et Chaire ECOSANTE, Université Cheikh Anta Diop (ACAD), Dakar, Sénégal

K. W. Fomba et al. (eds.), *Advances in Air Quality Research in Africa*,
Advances in Science, Technology & Innovation, https://doi.org/10.1007/978-3-031-53525-3_22

(Martin et al., 2013; Quansah et al., 2017). Hence, the project entitled Ecohealth Chair in Urban Air Pollution and Non-Communicable Respiratory Diseases had the overall objective of contributing to the improvement of the understanding of urban air pollution and its impacts on respiratory non-communicable diseases and putting in place community-based prevention strategies; this is to ensure a healthier, livable and sustainable urban environment. Given the multifactorial aspects of air pollution and the importance of the evolution of non-communicable diseases (NCDs), in particular respiratory diseases, the use of an approach integrating the different contexts, namely environmental, economic, and socio-cultural, is necessary (Martin et al., 2013; Webb et al., 2010). This Ecohealth chair implemented in 2015 at the University of Abomey-Calavi (Benin) is a regional initiative in West Africa covering mainly four countries, Benin, Burkina Faso, Ivory Coast, and Senegal.

1.1 Research Questions

1. What is the air quality in the region?
2. What are the environmental, health, socio-anthropogenic, and economic inequalities associated with urban air pollution?
3. What are the stumbling blocks and hindrances to the enforcement of regulations related to air pollution?

2 Methodology

In an "Ecohealth" approach that advocates interdisciplinarity, various research methods were used during this project, which included a research component focusing on environmental, health, and socio-economic aspects, and an action component involving populations and institutions. In the research component, studies were conducted to characterize and measure the health risks related to air pollution in various vulnerable and particularly exposed groups (children, women, motorcycle cabs, bus drivers, and shopkeepers living along the roads). Outdoor air pollution was studied in Cotonou and Dakar based on a principal marker that is the transport system. Indeed, the city of Dakar is characterized by a relatively well-organized public transport system and individual means of transport composed of four-wheel engine. In the city of Cotonou, urban transport is mainly individual and mostly consists of two-wheeled vehicles (four-stroke and two-stroke motorcycles). Due to financial constraints, in Abidjan and Ouagadougou, a comparative analysis of indoor pollution parameters was conducted. For the action component, an intervention called Air-sain has been implemented in each country and has helped, in addition to mass actions, to encourage the use of less polluting energy sources through their subsidy and to influence a thousand community and institutional actors for better air quality.

3 Results and Discussion

3.1 Level of Exposure

Indoor and outdoor pollution levels measured in urban West Africa are high with greater concentration at crossroads (Attoh Touré et al., 2018; Houngbégnon et al., 2019; Kouao et al., 2019c). In Benin, Houngbégnon et al. measured $PM_{2.5}$ concentrations along a high-traffic road in Cotonou (Table 1) (Houngbégnon et al., 2019). It was found that crossroads are the areas with significantly higher $PM_{2.5}$ exposure. In the rainy season, the mean $PM_{2.5}$ concentration was 463.25 ± 66.21 µg/m^3 at the crossroads and 264.75 ± 50.97 µg/m^3 along the roads. In the dry season, the average $PM_{2.5}$ concentration was 232.75 ± 97.29 µg/m^3 at the crossroads and 123.31 ± 63.79 µg/m^3 along the roads. In both dry and rainy seasons, the peak $PM_{2.5}$ concentrations were observed from 7:00 am to 9:00 am and from 7:00 pm to 9:00 pm.

Table 1 Comparison of $PM_{2.5}$ concentration between crossroads and outside crossroads and between season and place in Cotonou (Benin) (Houngbégnon et al., 2019)

	Coefficient	95% of IC	p-value	$PM_{2.5}$ concentration (µg/m^3)	
				Rainy season	Dry season
Places					
Intercept	114.61	[96.85; 132.36]	<0.001	–	–
Outside of crossroad	–	–	–	264.75 ± 50.97	123.31 ± 63.79
Crossroad	152.97	[123.38; 182.56]	<0.001	463.25 ± 66.21	232.75 ± 97.29
Seasons					
Rainy season	158.85	[135.18; 182.52]	<0.001	–	–
Dry season					

Kouao et al. in a study of air pollution exposure in children under 5 years of age in urban Abidjan found daily averages of $PM_{2.5}$ indoors (121 ± 12 µg/m³ in Andokoi and 32 ± 3 µg/m³ in Lubafrique) and outdoors (117 ± 8 µg/m³ in Andokoi and 41 ± 4 µg/m³ in Lubafrique) well above the World Health Organization's recommended value for air quality (25 µg/m³) (Table 2) (Kouao et al., 2019c). Also, the number of windows in bedrooms and the location of kitchens outdoors were negatively correlated with indoor $PM_{2.5}$ concentration. Outdoor $PM_{2.5}$ concentrations were higher with cooking fuel types, especially for wood and charcoal.

3.2 Risk Factors

The intensity of urban traffic, the use of biomass, the location of the kitchen inside the dwelling, the low number of windows in the rooms, and the use of products to control disease-carrying insects (anopheles, cockroaches, etc.) and to improve olfactory sensations (incense, air fresheners, etc.) have been identified as risk factors for air pollution (Attoh Touré et al., 2018; Houngbégnon et al., 2019; Kafando et al., 2018, 2019a, 2019b; Kouao, 2019; Kouao et al., 2019a; Kouao et al., 2019b; Ndong Ba et al., 2019; Sana et al., 2018). Outdoor and domestic pollution would promote in the short term the occurrence of cough, cold, and acute respiratory infections; in the medium term the occurrence of asthma, allergic rhinitis, and chronic obstructive pulmonary disease (COPD); in the long term the occurrence of lung cancer or a risk of genetic abnormalities (Kafando et al., 2018; Lawin et al., 2018a; Sylla et al., 2017a, 2017b, 2018). People working at crossroads are exposed to significantly more respiratory risks than those working along roads (Houngbégnon et al., 2020). Vulnerable groups such as children, women, and occupationally exposed individuals are particularly at risk because they do not have sufficient knowledge of the hazards involved (Lawin et al., 2018b; Ouédraogo et al., 2018).

Table 2 Comparison of $PM_{2.5}$ concentration in indoor and outdoor between study's sites in Abidjan (Ivory Coast) (Kouao et al., 2019c)

	$PM_{2.5}$ concentration (µg/m³)	
	Indoor	Outdoor
Sites		
Andokoi	121 ± 12	32 ± 3
Lubafrique	117 ± 8	41 ± 4

3.3 Level of Implementation of Regulations

Regarding the level of implementation of policies and regulations adopted to reduce air pollution, important mechanisms and institutions exist to stop environmental degradation and protect air quality. The interventions in this matter are based on the conventions and agreements signed between the States and the laws related to petroleum products and transport as well as the standard norms of pollutant emission. In the application, they encounter obstacles related to governance and resources made available. Benin has more regulatory texts than other countries and Senegal deploys more material resources for air quality control. In addition, some legal provisions appear to be outdated given the evolution of the phenomenon and others seem to be unknown to the actors. The instruments for regulating air pollution require capacity building and policies for air quality preservation in West Africa.

3.4 Implementation of Air-sain

The implementation of the Air-sain program in the four countries was based on the results of a KAP survey. Indeed, the KAP survey allowed us to establish an initial picture of the potential beneficiary populations and to better understand the intervention context. In general, the aim was to implement a strategy to raise awareness and support the acquisition of alternative tools through a participatory approach in the targeted neighborhoods and to contribute to the strengthening of regulations and good practices in industries and state institutions through a multi-sectoral approach. At the end of these intervention rounds, a post-intervention KAP survey was used to determine the final (post-intervention) level to assess the changes in the targets. Thus, the intervention was based on two components. The first was community-based, involving populations through awareness-raising and support for ecological cooking tools, and the second was institutional, targeting twenty ministries, strategic departments, decentralized authorities, and industry representatives. Regarding this last component, it should be noted that the representatives of each structure participated in about 10 meetings (quarterly and preparation meetings) during the implementation of the intervention. Similarly, specific meetings with key actors were held to facilitate the implementation of the intervention and ensure its success. In general, the intervention allowed the research teams to be solicited to participate in radio and television programs on air pollution, which made the

Chair more visible. In the same way, by the various activities, it is to note awareness of the actors at various levels on the health risks and the behaviors harmful to health and the environment. At the level of the institutional actors, the intervention will have served to take initiatives and actions, in particular the realization of cleanliness campaigns in the neighborhoods.

4 Conclusion

Air pollution is a serious problem in sub-Saharan Africa and the rest of the world. National legislation and institutional frameworks to ensure air quality appear to be ineffective in implementing the mitigation measures needed to reduce air pollutants in this region. The findings of this study urgently warrant the identification and implementation of effective organizations and initiatives that will address the increasing air pollution in the region. In general, the Ecohealth Chair project, through its different axes, has helped to raise local awareness of the air pollution problem and to initiate a change in the practices of communities and institutions.

References

Amegah, A. K., & Agyei-Mensah, S. (2017). Urban air pollution in Sub-Saharan Africa: Time for action. *Environmental Pollution, 220*(Pt A), 738–743. https://doi.org/10.1016/j.envpol.2016.09.042

Attoh Touré, H., Coulibaly, M., Kouao, R., Tiembré, I., Konaré, A., & Dagnan, N. S. (2018). Fine particulate matter pollution in African urban area: An urgent need for action. *International Journal of Advanced Research, 3*(1), 320–325.

Houngbégnon, P., Amadou, A., Lawin, H., Atindegla, E., Kêdoté, M., & Fayomi, B. (2020). Respiratory symptoms associated with workplaces located along a road with high-traffic at Cotonou Benin. *Open Journal of Respiratory Diseases, 10*, 32–42.

Houngbégnon, P., Ayivi-Vinz, G., Lawin, H., Houessionon, K., Tanimomon, F., Kêdoté, M. et al. (2019). Exposure to $PM_{2.5}$ related to road traffic: Comparison between crossroads and outside of crossroads at Cotonou, Benin. *Open Journal of Air Pollution, 8*, 108–117.

Kafando, B., Windinpsidi Savadogo, P., Millogo, T., Sana, A., Kouanda, S., & Sondo, B. (2018). Pollution de l'air intérieur et prévalence des infections respiratoires aiguës chez les enfants à Ouagadougou. *Santé Publique, 30*(4), 575–586.

Kafando, B., Windinpsidi Savadogo, P., Millogo, T., Sana, A., Sanon, S., Kouanda, S., & Sondo, B. (2019a). Exposure to conditions of high concentrations of indoor air pollutants and prevalence of ARIs in children under 5 years of age Ouagadougou/Burkina Faso. *Environmental Science and Public Health, 3*(3), 298–314.

Kafando, B., Savadogo, P. W., Sana, A., Bagnoa, V., Lompo, S., Kouanda, S., et al. (2019b). Pollution intérieure par les PM2,5 issues des combustibles utilisés pour la cuisson des repas et risques sanitaires dans la ville de Ouagadougou. *Environnement Risques et Santé, 18*, 245–253.

Katoto, P. D. M. C., Byamungu, L., Brand, A. S., Mokaya, J., Strijdom, H., Goswami, N., et al. (2019). Ambient air pollution and health in Sub-Saharan Africa: Current evidence, perspectives and a call to action. *Environmental Research, 173*, 174–188. https://doi.org/10.1016/j.envres.2019.03.029

Kouao, A. K. R. (2019). Prevalence of Asthma in children under 5 years old exposed to air pollution in Abidjan, (Côte D'ivoire). *International Journal of Recent Scientific Research, 10*(07), 33353–33358.

Kouao, A. K. R., Lognon, J. L., & Brou, N. (2019). Déterminants sociologiques de l'exposition des enfants de 0 à 5 ans à la pollution de l'air dans la commune de Yopougon. *Review of Social Economy, 16*, 21–38.

Kouao, A., N'datchoh, E., Yoboue, V., Silue, S., Attoh, H., Coulibaly, M., et al. (2019b). Exposure to indoor and outdoor air pollution among children under five years old in urban area. *Global Journal of Environmental Science and Management, 5*(2), 191–202.

Kouao, A., N'datchoh, E., Yoboue, V., Silue, S., Attoh, H., Coulibaly, M., et al. (2019c). Exposure to indoor and outdoor air pollution among children under five years old in urban area. *Global Journal of Environmental Science and Management, 5*(2), 191–202.

Lawin, H., Ayi Fanou, L., Hinson, V., Stolbrink, M., Houngbegnon, P., Kedote, N. M., et al. (2018a). Health risks associated with occupational exposure to ambient air pollution in commercial drivers: A systematic review. *International Journal of Environmental Research and Public Health, 15*(9), 2039. https://doi.org/10.3390/ijerph15092039

Lawin, H., Ayi Fanou, L., Hinson, V., Tollo, B., Fayomi, B., & Ouendo, E. M. (2018b). Facteurs de risques pro-fessionnels et perceptions de la pollution de l'air des taxis motoristes à Cotonou Bénin. *Santé Publique, 30*(1), 125–134.

Martin, W. J., 2nd., Glass, R. I., Araj, H., Balbus, J., Collins, F. S., Curtis, S., et al. (2013). Household air pollution in low- and middle-income countries: Health risks and research priorities. *PLoS Medicine, 10*(6), e1001455. https://doi.org/10.1371/journal.pmed.1001455

Ndong Ba, A., Verdin, A., Cazier, F., Garcon, G., Thomas, J., & Cabral, M. (2019). Individual exposure level following indoor and outdoor air pollution exposure in Dakar (Senegal). *Environmental Pollution, 248*, 397–407.

Odonkor, S. T., & Mahami, T. (2020). Knowledge, attitudes, and perceptions of air pollution in Accra, Ghana: A critical survey. *Journal of Environmental and Public Health*, 3657161. https://doi.org/10.1155/2020/3657161.

Ouédraogo, Z. G. M. E., Ouédraogo, V., & Drabo, M. K. (2018). Environnement réglementaire et système de contrôle de la filière menuiserie bois de la ville de Ouagadougou au Burkina Faso. *Science et Technique, Sciences de la Santé, 4*(1), 105–116.

Quansah, E., Katata, G., Mauder, M., Annor, T., Amekudzi, L. K., Bliefernicht, J., Heinzeller, D., Balogun, A. A., & Kunstmann, H. (2017). Numerical simulation of surface energy and water balances over a semiarid grassland ecosystem in the West African Savanna. *Advances in Meteorology, 2017*, Article ID 6258180. https://doi.org/10.1155/2017/6258180.

Sana, A., Somda, S. M. A., Meda, N., & Bouland, C. (2018). Chronic obstructive pulmonary disease associat-ed with biomass fuel use in women: A systematic review and meta-analysis. *BMJ Open Respiratory Research, 5*, e000246.

Sylla, F. K., Faye, A., Fall, M., & Tal-Dia, A. (2017a). Air pollution related to traffic and chronic respiratory diseases (Asthma and COPD) in Africa. *Health, 9*, 1378–1389.

Sylla, F. K., Faye, A., Fall, M., Lo, M., Diokhané, A., Touré, N. O., & Tal-Dia, A. (2017b). Near-road exposure to air pollution and allergic rhinitis: A cross-sectional study among vendors in Dakar

Senegal. *Occupational Diseases and Environmental Medicine, 5*, 106–120.

Sylla, F. K., Faye, A., Diaw, M., Fall, M., & Tal-Dia, A. (2018). Traffic air pollution and respiratory health: A cross-sectional study among bus drivers in Dakar (Senegal). *Open Journal of Epidemiology, 8*, 1–13.

Webb, J. C., Mergler, D., Parkes, M. W., Saint-Charles, J., Spiegel, J., Waltner-Toews, D., et al. (2010). Tools for thoughtful action: The role of ecosystem approaches to health in enhancing public health. *Canadian Journal of Public Health, 101*(6), 439–441. https://doi.org/10.1007/BF03403959

Policy, Regulation, Public Awareness and Materials for Sensing Devices

20 Years of Air Quality Policy in Morocco

Celia Mir-Alvarez, Soraya Boudia and Bertrand Lefebvre

Abstract

Urban air pollution in low- and middle-income countries (LMICs) is a growing global health concern. Despite this, research into issues surrounding air quality management in LMICs is lacking. Especially in Africa, it can be difficult to find even basic information on air quality legislation and monitoring networks, inhibiting further discussions into urban air quality governance. This study aims to shed light onto the status of air quality management in Morocco as part of a doctoral thesis. A particular focus was placed on Casablanca, the country's most populous city. Data were collected through literature review and qualitative interviews with 35 experts in Rabat and Casablanca, Morocco. Results show that Morocco has made notable progress in developing an air quality surveillance network through their National Air Quality Plan, but staffing and budgeting issues have hampered progress. Translating national policies to local contexts is also a concern. Finally, experts are divided when discussing transparency in sharing air quality data with the population. Further research is needed to better link the wider social and political context, as well as the public and global health spheres, to air quality governance in the country.

Keywords

Morocco · Urban air pollution · Air quality monitoring · Air quality governance · Public health · Global health

1 Introduction

Ambient air pollution is a global health threat causing over four million premature deaths each year. 90% of these occur in low- and middle-income countries (LMICs) (WHO, 2023); this is a particular concern in rapidly growing and industrializing LMIC cities (Globalsmog Homepage, 2023). Despite decades of research highlighting the adverse effects of PM_{10}, $PM_{2.5}$, and other pollutants on respiratory, cardiovascular, and even maternal health, it seems LMICs continue to struggle to make air quality management a priority (Globalsmog Homepage, 2023). There is very little research into stakeholders' views, perceptions on air pollution as a public health problem, conflicting agendas, or citizen awareness of air pollution in LMIC cities that could shed light onto the reasons for this struggle. Potential issues in the way international (health) agencies such as the World Health Organization (WHO) and the World Bank have interacted with LMICs with regards to air pollution have also not been sufficiently explored.

Especially for the African continent, it can even be difficult to obtain information on the status of a country or city's air quality legislation or monitoring network due to lacking research. Only a handful of reviews, published recently, have attempted to document this for a number of African countries and cities (Abera et al., 2021; Mir Alvarez et al., 2020; Schwela, 2012). Understanding the existing infrastructure, its evolution over the years, and the roles of stakeholders is an essential first step before diving deeper into more complex issues surrounding urban air quality governance.

This study aims to lay out such groundwork for the city of Casablanca, Morocco. Located in the Middle East and North Africa (MENA) region, Morocco's geographical location makes it particularly susceptible both to the effects of climate change and to natural and anthropogenic air pollution from nearby areas (Sekmoudi et al., 2022). A rapidly urbanizing population with ever-growing energy

C. Mir-Alvarez (✉) · S. Boudia
Université Paris Cité, CNRS, Inserm, Cermes3, 45 Rue Des Saints-Pères, 75270 Paris, France
e-mail: celia.mir-alvarez@u-paris.fr

B. Lefebvre
Institut Français de Pondychéry, 11, St Louis St., White Town, Puducherry 605001, India

K. W. Fomba et al. (eds.), *Advances in Air Quality Research in Africa*,
Advances in Science, Technology & Innovation, https://doi.org/10.1007/978-3-031-53525-3_23

needs and a strong dependence on coal for energy production greatly influence air quality in cities. Regarding public health, Morocco's burden of disease is shifting towards non-communicable diseases, but the country continues to struggle with infectious diseases like HIV and tuberculosis in some areas (Chadli et al., 2018). Casablanca was chosen as it is Morocco's largest city, hosting a population of 3.8 million (Casablanca Population, 2023) and 32% of its GDP (SEDD, 2018). It is also one of the world's most densely populated cities, with just one-tenth of the recommended green spaces per inhabitant (Casablanca, 2023; Casablanca World's Fifth Most Crowded City, 2023). A large port, an airport, strong industrial activity, and a large, dated vehicle fleet are just some of the pollution sources threatening the health of Casablanca's citizens.

Morocco's governmental configuration makes discussing air quality and health data a delicate matter. However, as its population continues to urbanize—about 75% of Morocco's population will live in cities by 2050 (Country Profiles, 2023)—it is important to shed some light onto the existing mechanisms used to protect the urban population from adverse health outcomes. The objective of this study is therefore to describe the evolution of air quality surveillance in the country, to mention relevant stakeholders, and to understand the place of public health in air quality management. This research constitutes preliminary findings of a PhD thesis in the field of Science and Technology Studies. Although funding is separate, this PhD thesis takes place under the Globalsmog project, a multidisciplinary project that aims to address the questions above for 10 additional LMIC cities across Africa and Asia (Globalsmog Homepage, 2023).

2 Methods

An in-depth literature and gray literature review were conducted between November 2021 and April 2022 to identify pollution sources, research groups, existing legislation and policies, and studies on health effects of air pollution in Morocco. With little to no publicly available air quality and health data, expert qualitative interviews were essential to fill in gaps identified through the literature review. A fieldwork visit to Rabat and Casablanca was carried out between May and June 2022 for this purpose. Interviews were semi-structured, were recorded only when consent was given, and lasted between 60 and 120 min. Interviews were anonymized, transcribed, and analyzed using NVivo, version 12.

A total of 35 experts were consulted and were cited in the results using anonymous identifiers. They included five members of Moroccan ministries (P16, P22, P29, P30, P35) four members or ex-members of NGOs or environmental activists (P8, P13, P25, P31), one medical doctor (P23), two independent environmental consultants (P12, P26), one elected representative in Casablanca (P18), three employees at foreign or international agencies (P1, P17, P19), four members of state agencies (P14, P15, P32, P33), 13 Moroccan or foreign researchers (P2–P7, P9–P11, P21, P27, P28, P34), one teacher (P24), and one employee at a private company (P20). Categories are intentionally broad to protect participants' identity.

3 Results and Discussion

Air quality has been monitored in Morocco since the early 2000s, after the Mohammed VI Foundation for the Protection of the Environment (FM6E) and the Government of Monaco jointly purchased three fixed air quality monitoring stations for Rabat and Casablanca [P12; P29]. The stations' data analysis and maintenance were entrusted to the DGM, Morocco's national meteorology agency, through a memorandum of understanding between the then Minister of Land Management, Water and the Environment and the DGM in 2007 [P29]. Over the years, FM6E, the DGM, and the Moroccan government gradually provided financing to purchase new stations [P12; P25; P32]. Today, there are 34 fixed automatic and 4 mobile stations located throughout the country, with 17 fixed stations in Casablanca and its vicinity (Ministère de la Transition, 2022). The DGM has continued to oversee the stations until very recently when the country's National Air Quality Plan (PNAir) established in 2018 that these responsibilities were to shift to the National Laboratory for the Surveillance and Study of Pollution. This transition is currently underway and is expected to conclude by the end of this year, but the DGM will continue to have data access for research purposes [P12; P15; P29; P35].

PNAir contemplates the expansion of the surveillance network to 140 stations by 2030 [P15]. These will be gradually acquired and funding will come from the Ministry of Energy Transition and Sustainable Development (MTEDD), the Ministry of the Interior, grants from partner countries like France, or Morocco's regional governments, depending on each case [P12; P29; P35]. Participants expressed concern that they did not yet know whether the size of the teams that will be tasked with the analysis and maintenance of stations will increase to meet the needs of the growing network; in the Casablanca-Settat region, for example, the regional delegation of MTEDD has a team of just eight people to manage the entirety of the region's environmental actions, and there are no air quality specialists [P16; P29; P32; P33]. Understaffing has been an issue for the DGM in the past, and led to the permanent damage of several air quality stations due to insufficient maintenance [P12; P32; P33]. Between 2016 and 2020, no air quality data was collected for the Casablanca-Settat region because stations were out of order [P29]. It is unclear how many stations are functioning correctly across the country today, although there are currently efforts in place to restore and replace malfunctioning stations. As services tasked with air quality management are on the lower end of the organizational structure of their respective ministries or agencies, raising concerns up to the levels of decision-makers is often difficult [P16; P29; P32; P33; P35].

Other pillars of PNAir include increasing citizen awareness, reducing polluting activity, and strengthening the legislative framework [P12; P30]. As of today, however, the Ministry of Finance has only guaranteed funding mechanisms for the acquisition of stations [P35]. Yet these other pillars also require action: Morocco has several laws establishing limits for ambient air quality, but others limiting vehicle or industrial emissions are lacking the decrees that make them applicable [14; P18; P19; P22]. Many experts agreed that laws limiting emissions are not sufficiently enforced [P2; P6-P8; P18; P19; P28; P34]. Some also argued that increasing vehicle inspections, for example, could contribute to both reduce emissions and raise citizen awareness, as they do not yet make the link between air quality and individual actions [P20; P27; P29; P31; P34]. Concerns about effectively implementing national legislation in a complex city like Casablanca arose during interviews as well [P9; P16; P18; P20; P21].

It is difficult to put citizens at fault for lacking awareness in a country where air quality data is not readily shared with the population. While some participants claimed that this was simply due to difficulties with monitoring stations in recent years [P29], and that they wanted more transparency in the future [P25; P26; P27], others cited "not wanting to alarm the population" and preferred to share data only if pollutant concentrations posed a serious risk for human health (there is a law setting an "alert threshold", after which authorities are required to inform the population) [P12; P14; P15]. Civil society leaders and environmental activists, as well as other researchers and experts, called for better communication of air quality data and stronger cooperation with local activists [P1; P13; P17; P24; P27; P31; P34]. Potential motivations for Morocco and other MENA countries to engage in more transparent, coherent behavior concerning air quality and environmental protection in general are under discussion [P10; 15; 16].

Difficulties in accessing data have also hampered academic research on air quality in Morocco. The few active research groups in the field engaged in a lengthy process to establish a convention with the DGM in order to access air quality data [P2-P5; P34]. This procedure, added to the low volume and quality of the data once received, could explain why such a small number of researchers outside the DGM research air quality in Morocco [P7; P27]. A lack of collaboration between actors and academia has resulted in the absence of updated emissions inventories and advanced predictive models that could serve to prepare for pollution episodes in advance. As of today, Moroccan authorities only find out about a high-pollution episode after it occurs [P29; P33].

Finally, there are no digital health records in Morocco that could show the impact of a pollution episode on the health system [P22; P23]. The few government-executed studies that link air quality and health effects were only possible through planned projects involving partnerships between the Ministry of Health, MTEDD, DGM, FM6E, and selected hospitals and health centers. Doctors from these centers were requested to record visits for illnesses associated with air pollution over a specific period of time. Two such studies have been carried out in Casablanca over the past two decades, and despite gaps in air quality data, found significant associations between black carbon and both mortality and hospital visits due to upper and lower respiratory tract infections in children under five years of age, among other findings [P22]. As a result of the study carried out in 2014, FM6E funded the development of a tool capable of carrying out real-time air quality and health data analyses. A service at the Ministry of Health has access to this tool, but without data, it cannot yet be put to use [P22; P25].

4 Conclusion

The present study has briefly summarized the status and evolution of air quality monitoring, data, and policy in Morocco over the past 20 years. Morocco has made notable progress and the country's future plans show promise. More research into the broader social, economic, and political context in Morocco that may influence decisions that impact air quality is needed. Further, a second fieldwork visit will be carried out to better understand the role of the public health community, the constraints of implementing PNAir in Casablanca, the influence of international agencies, and to observe whether the changes mentioned by interviewees really take place in the near future. These findings will be compared with other LMIC cities included in the Globalsmog project.

References

Abera, A., Friberg, J., Isaxon, C., Jerrett, M., Malmqvist, E., Sjöström, C., Taj, T., & Vargas, A. M. (2021). Air Quality in Africa: Public Health Implications. *Annual Review of Public Health, 42*, 193–210.

Ambient (outdoor) air pollution, World Health Organization. Retrieved May 8, 2023, from https://www.who.int/news-room/fact-sheets/detail/ambient-(outdoor)-air-quality-and-health.

Casablanca, «le cœur battant» éprouvé par la pollution, MAP - Agence Marocain de Presse. Retrieved May 8, 2023, from http://mapecology.ma/actualites/casablanca-coeur-battant-eprouve-pollution/.

Casablanca Population. (2023). World Population Review. Retrieved May 8, 2023, from https://worldpopulationreview.com/world-cities/casablanca-population.

Casablanca World's Fifth Most Crowded City: World Economic Forum, Morocco World News. Retrieved May 8, 2023, from https://www.moroccoworldnews.com/2017/05/217659/casablanca-worlds-fifth-most-crowded-city-world-economic-forum-morocco-wef.

Chadli, S., Taqarort, N., el Houate, B., & Oulkheir, S. (2018). Transition épidémiologique au Maroc (1960–2015). *Médecine et Santé Tropicales, 28*(2).

Country Profiles, United Nations World Urbanization Prospects. Retrieved May 8, 2023, from https://population.un.org/wup/country-profiles/.

Globalsmog Homepage. Retrieved May 8, 2023, from https://www.globalsmog.org/.

Mir Alvarez, C., Hourcade, R., Lefebvre, B., & Pilot, E. (2020). A scoping review on air quality monitoring, policy and health in West African cities. *International Journal of Environmental Research and Public Health, 19*(23), 1–28.

Schwela, D. (2012). *Review of urban air quality in Sub-Saharan Africa Region*. World Bank.

SEDD. (2018). Étude relative à la surveillance de la qualité de l'air. Mission 1: Rapport régional sur la qualité de l'air - région Casablanca-Settat. Ministère de l'Énergie, des Mines et du Développement Durable.

Sekmoudi, I., Bouakline, O., Khomsi, K., Idrissi, L., El Merabet, Y., Souhaili, Z., & El Jarmouni, M. (2022). A review on climate, air pollution, and health in North Africa. *Current Environmental Health Reports, 9*(2), 276–298.

Ministère de la Transition Énergétique et du Développement Durable: Surveillance de la Qualité de l'Air au Maroc: Rapport National (2022).

Production and Circulation of Local Knowledge About Air Pollution and Health Effects in Ghana

Jessica Pourraz

Abstract

According to the World Health Organization, air pollution is the main environmental health risk at the global scale, but it affects more particularly and severely populations of Global South cities, where environmental regulations remain weak. Data play an important role to shed light on the problem of air pollution and to trigger political responses. In this paper, I propose to analyze the conditions of production and circulation of data and local knowledge on air pollution and effects on health in Ghana. More specifically, I will focus on the actors and the resources they mobilize, what knowledge do they produce and how are they circulate and are used. This contribution is drawn from an ongoing (2021–) post-doctoral research project in sociology conducted in Accra and funded by the Institut Francilien Recherche Innovation en Société. It is based on a qualitative methodology made of observation at the Environmental Protection Agency (EPA-Ghana) and semi-structured interviews conducted in Accra ($N=35$) between August 2021 and May 2022 with academic and institutional actors. Preliminary results show that Ghana's air quality and health effects knowledge result from transnational financial, human and technological circulations. Overall knowledge produced remains limited to raw air quality (PM) data and lack of enough data in source apportionment and health effects. Moreover, these data are not disseminated to the public and seem to be partly ignored by the political authorities, enabling the formulation of mitigation policies.

Keywords

Ghana · Air pollution · Knowledge production and dissemination

1 Introduction

According to the World Health Organization (WHO), air pollution—ambient and indoor—is the main environmental risk to health. It causes 7 million deaths per year (WHO, homepage), due to respiratory infections such as Chronic Obstructive Pulmonary Disease (COPD), lung cancer and cardiovascular disease. It affects more particularly and severely the populations in the Global South where nearly 92% of deaths occur, as well as the poorest, reflecting significant social inequalities (Landrigan et al., 2017). WHO warns that COPD will become the third leading cause of death in 2030 with 90% of deaths occurring in the Global South. Particulate matter ($PM_{2.5}$ and PM_{10}),[1] which has adverse effects on human health, is the most commonly used measure of air pollution. WHO recommends $PM_{2.5}$ ambient concentration not to exceed 15 µg per cubic meter of air (µg/m^3) over a 24-h period and PM_{10} not to exceed 45 µg/m^3 (World Health Organization, 2021). Accra in Ghana presents PM levels well above WHO recommendations (Alli et al., 2021). In most African cities such as Accra, air pollution results from a combination of several emission sources such as vehicles, dust from roads and construction sites, combustion of biomass (charcoal, wood) and waste burning.

J. Pourraz (✉)
Centre Emile Durkheim, Sciences Po Bordeaux, 33600 Pessac, France
e-mail: jessica.pourraz@ird.fr

[1] PM_{10} includes particles with a diameter of less than 10 µm (micrometers), whereas $PM_{2.5}$ are those smaller than 2.5 µm and they progress more deeply in the respiratory system.

K. W. Fomba et al. (eds.), *Advances in Air Quality Research in Africa*,
Advances in Science, Technology & Innovation, https://doi.org/10.1007/978-3-031-53525-3_24

However, despite the scale and severity of the problem of air pollution, environmental regulations in Ghana remain weak. Whereas Environmental Protection Agency (EPA-Ghana) experts and local academic actors produce knowledge about air pollution since the beginning of 2000s, these data are not disseminated to the public and seem to be partly ignored by the political authorities, enabling the formulation of mitigation policies.

In this paper, I propose to analyze the use of scientific data about air pollution in Ghana and how they are translated, or not, into policies. To do so, I analyze the local conditions of production and circulation of data and knowledge by looking at the actors (State, academics, experts), their resources, what kind of knowledge do they produce and how do they circulate and are used?

2 Methods

This contribution is drawn from an ongoing (2021–) postdoctoral research project in sociology conducted in Accra and funded by the Institut Francilien Recherche Innovation en Société (IFRIS). It is based on a qualitative methodology made of observation and semi-structured interviews conducted in Accra ($N=35$) between August 2021 and May 2022 with academics, air quality experts, members of NGOs (Clean Air One Atmosphere, GreenAd) and WHO, engineers working for start-ups, medical practitioners, civil servants from the Accra municipality, the Ministries of Transport and Environment, the Ministry of Health and the National Petroleum Authority. Direct and participant observations were also carried out at the Air Quality Research Laboratory from the Department of Physics at the University of Ghana (UG), at the laboratory of the Environmental Quality Department of the EPA-Ghana and during their air quality monitoring tour of the Accra metropolis. This ethnographic work is complemented by the analysis of acts, EPA Ghana activity reports, press articles and social media posts.

3 Results

3.1 Health Effects and Air Quality Monitoring Data Capabilities in Ghana

Data collected during my fieldwork in Accra have allowed highlighting that the Environmental Protection Agency of Ghana (EPA-Ghana), under the Ministry of Environment, started air quality monitoring in 1997 as part of a program funded by the World Bank. However, air quality monitoring capabilities are only concentrated in Accra. Today the Accra metropolis has fifteen monitoring sites, of which only eight are operational due to lack of resources. Most of the monitoring sites are located along the main roads and are equipped with outdated mobile devices called Mini Vol, which provide $PM_{2.5}$ or PM_{10}/m^3 concentration every 6 days for a period of 24 h. Only two sites are equipped with reference grade monitors, which produce continuous and real-time PM data using scattered light spectrometry for measurement. They are considered as reference monitors, worth more than 100 K€ each. They were purchased in 2021 under the World Bank's Pollution Management and Environmental Health Program funded by the governments of Norway, Germany and the UK. They are set up respectively on the campus of the University of Ghana in the Physics Department and in the compound of an elementary school in the Adabraka neighborhood, not far from the Agbogbloshie landfill site, from which significant toxic fumes are emitted due to the burning of electronic waste (Feldt et al., 2014).

Because of their limited resource, it's not possible for a country such as Ghana to deploy reference grade monitors widely. Therefore, low-cost sensors (LCS) have become very popular among academic actors, citizens, NGOs, activists and even regulators because it allows producing air quality data at lower costs in places where data are sparse. Although some experts may question the validity of the PM measurements produced by LCS, they constitute alternative or even complementary networks to the institutional surveillance system.

In Accra, scientists from the Departments of Physics and Geography at the University of Ghana have had academic partnerships since 2006 with colleagues at the Harvard School of Public Health and more lately the Public Health School from the Imperial College in London. Through these partnerships, they have been able to produce data on PM and other gases in different areas of Accra based on income levels of the population (Arku et al., 2008; Dionisio et al., 2010; Rooney et al., 2012). Through "Global North-Africa" academic partnerships, these researchers could get LCS imported from the United States (US). On the other hand, EPA Ghana experts have also received LCS from their colleagues working at the US EPA.

Both experts from the EPA Ghana and local academics produce also knowledge on source apportionment, but at a very limited scale (EPA Ghana, 2018). Indeed, to do so it requires gravimetric analyses and EPA Ghana laboratory staff can't regularly carry out these tests due to lack of financial means, reference standards and proper instruments. This constitutes a huge gap because the production of causal knowledge—i.e. the origin and source of pollution—is instrumental to design mitigation policies.

Moreover, air quality control and the issue of air pollution in Ghana are under the responsibility of the Ministry of the Environment and my observation has highlighted the lack of leadership of the health sector in this matter. Yet academic works (Boutaric & Lascoumes, 2008) have shown how epidemiological investigations have been decisive in the past to define air pollution as a public health problem and instigate policy change. And so far, only one randomized control trial was conducted in 2011 by the Kintampo Health Research Centre from the Ghana Health Services in collaboration with Columbia Public Health in New York. This study has shown the negative effects of indoor pollution on newborn (Jack et al., 2015) and was continued since 2020 through the Child Lung Function study.

3.2 Circulation of Air Quality and Health Data in Accra

Data collected on the field and their analysis show that air quality data produced by EPA Ghana experts and academics do not circulate outside their own epistemological and professional communities. They are not disseminated to the public who have no knowledge of the pollution levels and the risks for their health.

While LCS are designed to be integrated into digital infrastructures, that facilitate the circulation, processing and formatting of data via Apps, websites, we do not observe any such digital platforms in Ghana at the moment (although this is in the pipeline according to my interlocutors in Accra). Moreover, there are still not enough public awareness and social mobilization to shed light on this issue.

Therefore, data generated in Ghana are primarily used to draft policy documents such as "The Greater Accra Metropolitan Areas Air Quality Management Plan" published in 2018 by EPA-Ghana, supported by the US-EPA. The plan, which has not been implemented yet, is the first of its kind in Ghana. Its main objective is to define mitigation strategies, primarily focusing on vehicles, power generation units and industries. But more importantly, the plan identifies gaps and weaknesses in EPA-Ghana's capacity in air quality monitoring and laboratory chemical analysis and calls for developing a comprehensive emissions inventory for the Greater Accra region (EPA Ghana, 2018).

In regards to the health effects, although the first clinical trial conducted in Kintampo has shown clear evidences on damages for health, the Occupational and Environmental Health Directorate from Ghana Health Services in Accra is still calling for a larger cohort epidemiological study to be conducted to demonstrate causality between air pollution and health effects.

At the level of health facilities, despite the existence of the District Health Information Management System (DHIMS) deployed in 2012 by the Ministry of Health, health data of ailments derived from various sources including air pollution do not provide the true reflection of respiratory incidences (EPA Ghana, 2018). Physicians themselves are not trained properly to identify air pollution as a risk factor, which tends to underestimate the health effects of air pollution such as COPB.

4 Conclusion

In this contribution, I have shown that Ghana's air quality knowledge and health data result from transnational financial, human and technological circulations. Ghana EPA experts and Ghanaian academics rely highly on external funding, mainly from the US and UK, to produce these data, highlighting the disengagement of the Government of Ghana in this area.

However, overall knowledge remains limited mainly to raw air quality (PM) data and lack of enough knowledge about the air pollution sources and the epidemiology. Moreover, data do not circulate to the public and remain between researchers and EPA experts. This problem is completely invisible letting people ignorance and there are no social network movements to take up the issue and put it on the political agenda, neither patient enough aware of this risk factor to alert on this public health issue.

EPA experts and colleagues from different ministries (industry, transport, energy, etc.) use the little data that does circulate to draft "The Greater Accra Metropolitan Areas Air Quality Management Plan". However, there are very few knowledge on sources apportionment, which doesn't allow to design proper mitigation policies. The Government of Ghana is asking for more health and pollution sources data to act and we observe a contradiction between the ever-increasing demand for data and the lack of means available to produce this knowledge.

Finally, the control of the Ministry of the Environment on the problem of air pollution and the lack of leadership of the health sector in this matter has effects on the framing of air pollution, not a public health problem, but as an environmental problem which has consequences on the policy responses proposed.

References

Alli, A., Clark, S, Hughes, A., Nimo, J., Bedford-Moses, J., Baah, S., Wang, J., Vallarino, J., Agyemang E., Barratt B., Beddows A., Kelly F., Owusu G., Baumgartner, J., Brauer, M., Ezzati, M., Agyei-Mensah, S. & Arku, R. (2021). Spatial-temporal patterns of ambient fine particulate matter ($PM_{2.5}$) and black carbon (BC) pollution in Accra. *Environmental Research Letters*, 16074013:3.

Arku, R., Vallarino, J., Dionisio, K., Willis, R., Choi, H., Wilson, J., Hemphill, C., Agyei-Mensah, S., Spengler, J. D., & Ezzati, M. (2008). Characterizing air pollution in two low-income neighborhoods in Accra Ghana. *Science of the Total Environment, 402*(2), 217–231.

Boutaric, F., & Lascoumes, P. (2008). L'épidémiologie environnementale entre science et politiques. Les enjeux de la pollution atmosphérique en France. *Sciences Sociales et Santé, 26*, 5–38.

Dionisio, K. L., Arku, R. E., Hughes, A. F., Vallarino, J., Carmichael, H., Spengler, J. D., Agyei-Mensah, S., & Ezzati, M. (2010). Air pollution in Accra neighborhoods: Spatial, socioeconomic, and temporal patterns. *Environmental Science & Technology, 44*(7), 2270–2276.

EPA Ghana. (2018). *The greater Accra metropolitan areas air quality management plan.*

Feldt, O., Fobil, J., Wittsiepe, J., Wilhelm, M., Till, H., Zoufaly, A., Burchard, G., & Göen, T. (2014). High levels of PAH-metabolites in urine of e-waste recycling workers from Agbogbloshie Ghana. *Science of the Total Environment, 466–467*, 369–376.

Jack, D., Asante, K., Wylie, B., Chillrud, S., Whyatt, R., Ae-Ngibise, K., Quinn, A., Yawson, A., Boamah, E., Agyei, O., Mujtaba, M., Kaali, S., Kinney, P., & Owusu-Agyei, S. (2015). Ghana randomized air pollution and health study (GRAPHS): Study protocol for a randomized controlled trial. *Trials, 16*, 420.

Landrigan, P. J., et al. (2017). The Lancet commission on pollution and health. *The Lancet, 391*(10119), 462–512.

Rooney, M. S., Arku, R. E., Dionisio, K. L., Paciorek, C., Friedman, A. B., Carmichael, H., Zhou, Z., Hughes, A. F., Vallarino, J., Agyei-Mensah, S., Spengler, J. D., & Ezzati, M. (2012). Spatial patterns of particulate matter sources and pollution in four communities in Accra Ghana. *Science of the Total Environment, 435–436*, 107–114.

WHO Homepage. Retrieved February 15, 2022 from, https://www.who.int/data/gho/data/themes/air-pollution?lang=en.

World Health Organization. (2021). WHO global air quality guidelines. Particulate matter ($PM_{2.5}$ and PM_{10}), ozone, nitrogen dioxide, sulfur dioxide and carbon monoxide. Geneva.

Effects of Doping on the Gas-Sensing Response of Ga-Doped ZnO Nanoparticles

Marwa El Beji, Soumaya Jaballah, Mohamadou Ba, Nourredine Bouguila, Riadh Souissi, Brahim Bouricha, Hassen Dahman and Lassaad El Mir

Abstract

This research depicted a modified sol–gel method to synthesize nanoparticles of gallium-doped zinc oxide (ZnO) with the intention of creating gas sensor devices that had enhanced sensing layers. Various techniques were employed to characterize these nanoparticles, including X-ray diffraction (XRD) for structural analysis, transmission electron microscopy (TEM) and scanning electron microscopy (SEM) for morphological examination. The XRD results confirmed the good crystallinity of the samples in the hexagonal system. Both SEM and TEM images revealed that the nanoparticles had a size of approximately 30 nm, and the incorporation of gallium as a dopant reduced their size. Following that, gas sensors based on gallium-doped ZnO were tested at temperatures ranging from 250 to 450 °C. According to the experimental results, ZnO doped with 3% gallium exhibited heightened response when exposed to 500 ppm ethanol at its operating temperature of 450 °C. In comparison to methanol gas, the response and recovery times were shorter, and the selectivity was greater. These outcomes have significant implications for controlling air quality, particularly in industries that utilize organic solvents.

Keywords

Zinc oxide nanoparticles · Gallium · Sol–gel process · Physical properties · Ethanol · Gas sensors

1 Introduction

Air pollution attributed mainly to automobile exhaust and factory emissions has become a threat to the environment. Flammable and explosive gases can cause loss of life and environmental damage. The use of semiconducting metal oxides to efficiently detect different harmful and polluting gases with high sensitivity and selectivity has been researched for a while as part of attempts to enhance air quality (Talazac et al., 2004). Zinc oxide is capable of detecting various gases such as CO (Mahajan & Jagtap, 2020), ethanol (Huang et al., 2018), NH_3 (Nakarungsee et al., 2020) and toluene (Liu et al., 2022). Doping ZnO with III-A elements group such as Al, In, and Ga has been investigated by many researchers. The doped material improves electrical conductivity and encourages gas-sensing properties (Dhahri et al., 2015; Mahmoud et al., 2019). Yu et al. (2015) posted that Sn-Ga co-doped ZnO exhibited superior detection performance for ethanol at 225 °C in comparison with ZnO nondoped. Sun et al. (2015) reported that the detection limit of Er-doped ZnO nanofibers for ethanol was around 200 ppm, showing the best response to 200 ppm ethanol was 37.5 at 240 °C. In addition, Cao et al. (2018) elaborated Aluminium doped ZnO nanosheets, they have reported that ethanol sensing test reach the highest response (90.2) at optimal temperature of 370 °C. We study in this paper the performance of low-cost produced ZnO-based sensor with the aim of developing promising ethanol gas sensor.

M. E. Beji (✉) · S. Jaballah · M. Ba · N. Bouguila · H. Dahman · L. E. Mir
Laboratory of Physics of Materials and Nanomaterials Applied at Environment (LaPhyMNE), Faculty of Sciences in Gabes, Gabes University, 6072 Gabes, Tunisia
e-mail: marwaelbeji38@gmail.com

R. Souissi · B. Bouricha
Laboratory of Materials, Molecules and Applications (LMMA), El Marsa, Tunis, Tunisia

K. W. Fomba et al. (eds.), *Advances in Air Quality Research in Africa*,
Advances in Science, Technology & Innovation, https://doi.org/10.1007/978-3-031-53525-3_25

2 Experimental

2.1 Synthesis

In this work, we used a homemade sol–gel technique with supercritical drying of ethanol to synthesize pure and Ga-doped zinc oxide NPs according to protocol developed by El Mir and co-workers (El Mir et al., 2007, 2008). ZnO nanopowders were prepared by the dissolution of 16 g of $C_4H_6O_4Zn.2H_2O$ (Zinc acetate dehydrate-sigma-Aldrich 99–102%) in 112 ml of methanol under magnetic stirring. After 10 min, 188 ml of ethanol was introduced into the mixture, and the resulting homogeneous solution was poured into an autoclave. Thermal heating was then applied to the system until it reached the hypercritical conditions of ethanol, (Tc = 243 °C, Pc = 63.6) bar. The drying process extracts the solvent from the pores of the gel. The samples were finally annealed at 450 °C for 2 h under air into Nabertherm muffle furnace. Figure 1a illustrates a schematic representation of the used protocol. The samples will be labeled as G1ZO for 1% Ga-doped ZnO, G3ZO for 3% Ga-doped ZnO and G5ZO for 5% Ga-doped ZnO.

2.2 Characterization Techniques

X-Ray Diffractometer (XRD) Bruker AXS D8 Advance was used to study the microstructure of the samples using the CuK1 wavelength (0.15405 nm). Scanning electron microscopy (SEM) is a typical tool for producing micron-scale pictures of material morphology. The radiation is released when the sample interacts with X-rays. We have also investigated the morphology of the samples with transmission electron microscope (Philips PW1710) operating at 200 kV.

2.3 Sensing

The Ga-doped ZnO (GZO) samples were examined using the set-up illustrated in Fig. 1b. There are two sub-circuits in the global gas circuit. The organic vapor is initially tested using two mass flow controllers D1 and D2. These latter are responsible of dry airflow and organic vapor concentration adjusting. Temperatures ranging from 250 °C to 450 °C were used to conduct the tests. The sensor is activated by a light source (supplied by a stable generator) and controlled by a (HP 34401A) multimeter regulator, allowing the temperature of the semiconductor to be set by measuring the current across the RTD PT100 platinum probe.

3 Results and Discussion

We present in Fig. 2a the X-ray diffraction (XRD) pattern of GZO nanoparticles calcined at 450ºC for 2 h under air. The strong intensity and clarity of the diffraction peaks show that the samples are highly crystalline. No characteristic signal associated with gallium oxide is observed, which proves that Ga dopant was successfully incorporated into the structure.

SEM technique has been employed to investigate the morphological properties of G1ZO, G3ZO and G5ZO (Fig. 2b and c). They show that the powder involves nanoscale-sized finely crystallized particles. The morphology of the nanopowders' surfaces is extremely uniform.

Figure 3a–c shows TEM (transmission electron microscope) images of G1ZO, G3ZO and G5ZO annealed at 450 °C for 2 h under air. They are very fine with spherical shape, but the size varies between 30 and 40 nm. This proves that almost every grain is composed of a single crystallite and therefore no agglomeration occurs.

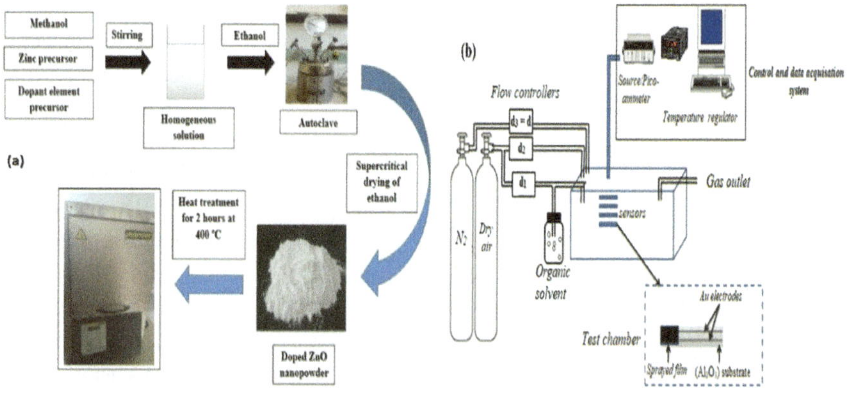

Fig. 1 **a** General steps of preparation of GZO samples through sol–gel process, **b** an illustration of gas sensing setup

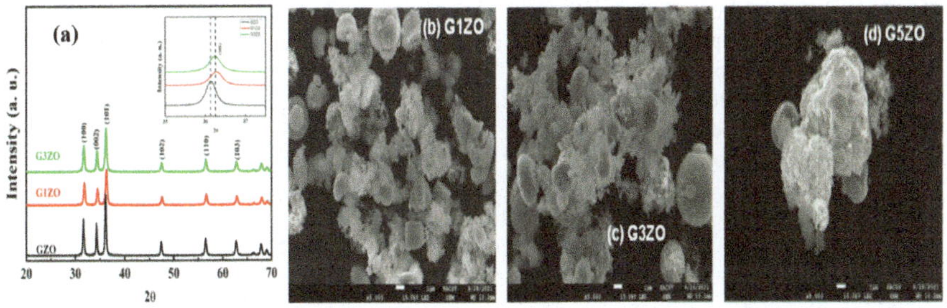

Fig. 2 **a** X-ray diffractograms of ZnO nanoparticles calcined at 450 °C. SEM images of nanoparticles of **b** G1ZO, **c** G3ZO and **d** G5ZO

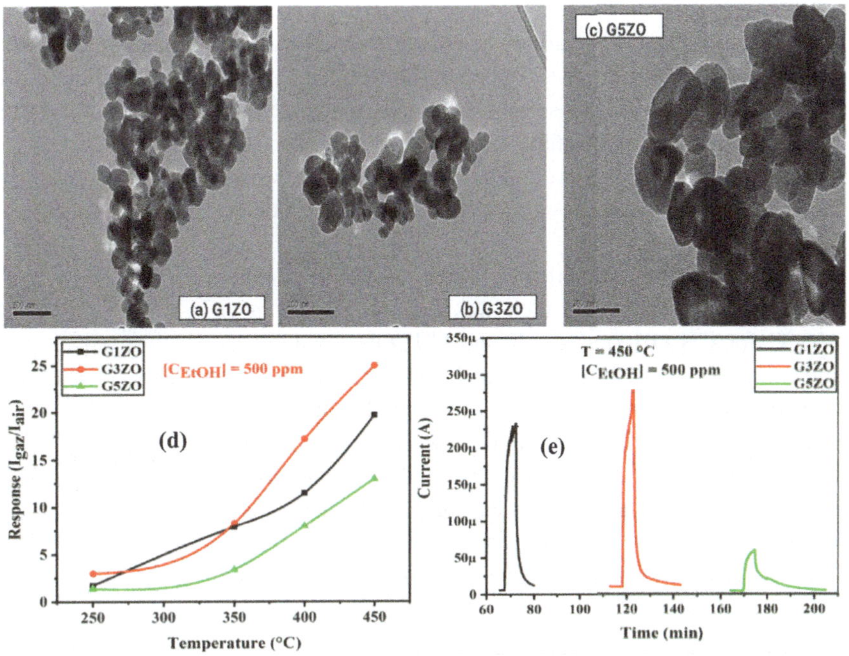

Fig. 3 TEM images for **a** G1ZO, **b** G3ZO and **c** G5ZO treated at 400 °C for 2 h under air. Response of GZO sensors to 500 ppm ethanol vapor **d** using various temperatures and **e** at 450 °C

Figure 3d illustrates the Ga-doped ZnO-based sensors' response towards 500 ppm ethanol at various temperature ranges (250–450 °C). The responses towards ethanol of all sensors are increasing with temperature elevating. The maximum response to 500 ppm ethanol was reached at 450 °C. The sensor employing the G3ZO sensor responds faster to 500 ppm ethanol and has a greater response value than the G1ZO and G5ZO sensors at the optimal operating temperature (450 °C) as shown in Fig. 3e. This temperature behavior response must be attributed to chemisorption and surface reactions.

Table 1 gives the different comparison data. The comparison results display that the G3ZO-based sensor has promising and competitive characteristics.

Figure 4a and b presents the response and recovery time of all samples (G1ZO, G3ZO, and G5ZO) when exposed

Table 1 Ga-doped ZnO gas sensing performance compared to existing literature

Samples	Signal type	Ethanol concentration (ppm)	Test temperature	Response	References
Ag/ZnO nanowires	Voltage	1000	25 °C	$R = (V_a - V_g)/V_a \times 100 = 90$	Zhang et al. (2021)
Ga-doped ZnO film	Resistance	250	450 °C	$R_a/R_g = 22$	Ramola et al. (2021)
Al-doped ZnO microrods	Conductance	8100	150 °C	$R_a/R_g = 231$	Sinha et al. (2020)
Ga-doped ZnO nanoparticles	Voltage	500	450 °C	$I_{EtOH}/I_{Dryair} = 25$	This work

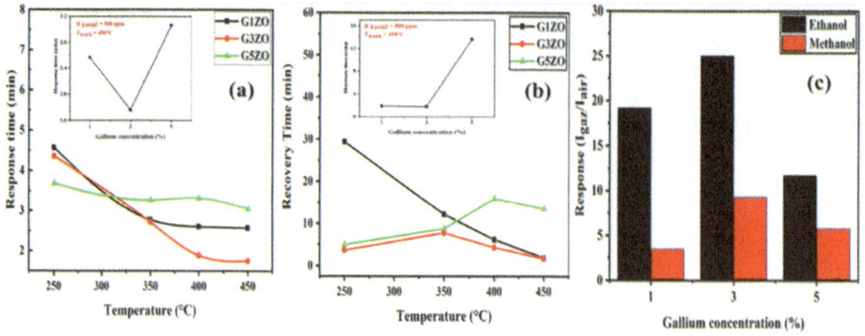

Fig. 4 Variation of the sensor parameters versus temperature when exposed to 500 ppm (**a**) response time and **b** recovery time. **c** Response of the samples for various gases (exposed to 500 ppm) at 450 °C

to 500 ppm ethanol at various temperatures, as well as their performance at the ideal operating temperature, it was observed that the G3ZO sensor exhibited a notable initial reduction in response time compared to the other sensors. Furthermore, it was found that the G3ZO sensor demonstrates faster recovery times compared to the remaining sensors.

Consider another important issue to take into account is the sensor selectivity, which was tested at 500 ppm at 450 °C for CH_3OH (methanol) and C_2H_5OH (ethanol) as shown in Fig. 4c. Both of G1ZO and G3ZO sensors exhibit higher response towards ethanol than methanol. As the size of the particle decreases, the material's specific surface area increases, there is a corresponding enhancement in gas adsorption. The detection capability of the gas sensor is enhanced, leading to improved response characteristics. XRD results prove that the G3ZO exhibits higher specific surface areas and lattice defects, which improves its detection capabilities.

4 Conclusion

Ga-doped ZnO NPs fabricated by sol–gel method and post-treated at 450 °C are investigated for ethanol detection. The microstructure and composition were verified through the results obtained from XRD, SEM and TEM analyses. XRD analysis of all the samples indicated the existence of sharp diffraction peaks of hexagonal symmetry of ZnO with no secondary phase. SEM and TEM images revealed a nanometer range size of the samples (about 30 nm). The continuation addition of gallium amount helps to reduce the nanoparticles' size. The gas sensing test proves that the GZO nanoparticle can be employed into ethanol detection. Outstanding gas detecting performances have been revealed for GZO samples. The ethanol vapor (500 ppm) shows good selectivity, high sensitivity and faster response (~1.9 min) and recovery time (~2.5 min). This pattern are due essentially to the small grain size and the high defect concentration which we enhance the specific surface area. The obtained results confirm that this material is promising for the control of air quality.

References

Cao, F., Li, C., Li, M., Li, H., Huang, X., & Yang, B. (2018). Direct growth of Al-doped ZnO ultrathin nanosheets on electrode for ethanol gas sensor application. *Applied Surface Science, 447*, 173–181.

Dhahri, R., Hjiri, M., Mir, L. E., Bonavita, A., Iannazzo, D., Leonardi, S. G., & Neri, G. (2015). CO sensing properties under UV radiation of Ga-doped ZnO nanopowders. *Applied Surface Science, 355*, 1321–1326.

El Mir, L., El Ghoul, J., Alaya, S., Ben Salem, M., Barthou, C., & Von Bardeleben, H. J. (2008). Synthesis and luminescence properties of vanadium-doped nanosized zinc oxide aerogel. *Physica B: Condensed Matter, 403*, 1770–1774.

Huang, Q., Navale, S. T., Cao, P., Han, S., Liu, X., Liu, W., Xu, W., Zhu, D., Stadler, F. J., & Lu, Y. (2018). Ultra-sensitive C_2H_5OH sensing properties of template-free solvothermal-processed hierarchical ZnO nanospheres. *Nanoscience and Nanotechnology Letters, 10*, 1651–1661.

Liu, H., Wang, Z., Cao, G., Pan, G., Yang, X., Qiu, M., Sun, C., Shao, J., Li, Z., & Zhang, H. (2022). Construction of hollow NiO/ZnO p-n heterostructure for ultrahigh performance toluene gas sensor. *Materials Science in Semiconductor Processing, 141*, 106435.

Mahajan, S., & Jagtap, S. (2020). Metal-oxide semiconductors for carbon monoxide (Co) gas sensing: A review. *Applied Materials Today, 18*, 100483.

Mahmoud, A., Echabaane, M., Omri, K., El Mir, L., & Ben Chaabane, R. (2019). Development of an impedimetric non enzymatic sensor based on ZnO and Cu doped ZnO nanoparticles for the detection of glucose. *Journal of Alloys and Compounds, 786*, 960–968.

El Mir, L., Ayadi, Z. Ben, Saadoun, M., Djessas, K., Von Bardeleben, H. J., & Alaya, S. (2007). Preparation and characterization of n-type conductive (Al, Co) co-doped ZnO thin films deposited by sputtering from aerogel nanopowders. *Applied Surface Science, 254*, 570–573.

Nakarungsee, P., Srirattanapibul, S., Issro, C., Tang, I. M., & Thongmee, S. (2020). High performance Cr doped ZnO by UV for NH_3 gas sensor. *Sensors and Actuators A: Physical, 314*, 112230.

Ramola, R. C., Negi, S., Rawat, M., Singh, R. C., & Singh, F. (2021). Annealing effects on gas sensing response of Ga-doped ZnO thin films. *ACS Omega, 6*, 11660–11668.

Sinha, M., Mahapatra, R., Mondal, M. K., Krishnamurthy, S., & Ghosh, R. (2020). Fast response and low temperature sensing of acetone and ethanol using Al-doped ZnO microrods. *Physica E: Low-dimensional Systems and Nanostructures, 118*, 113868.

Sun, Y., Zhao, Z., Li, P., Li, G., Chen, Y., Zhang, W., & Hu, J. (2015). Er-doped ZnO nanofibers for high sensibility detection of ethanol. *Applied Surface Science, 356*, 73–80.

Talazac, L., Barbarin, F., Mazet, L., & Varenne, C. (2004). Improvement in sensitivity and selectivity of InP-based gas sensors: Pseudo-schottky diodes with palladium metallizations. *IEEE Sensors Journal, 4*, 45–51.

Yu, L., Liu, S., Yang, B., Wei, J., Lei, M., & Fan, X. (2015). Sn-Ga co-doped ZnO nanobelts fabricated by thermal evaporation and application to ethanol gas sensors. *Materials Letters, 141*, 79–82.

Zhang, X., Wang, W., Zhang, D., Mi, Q., & Yu, S. (2021). Self-powered ethanol gas sensor based on the piezoelectric Ag/ZnO nanowire arrays at room temperature. *Journal of Materials Science: Materials in Electronics, 32*, 7739–7750.

SPRINGER NATURE

GPSR Compliance

The European Union's (EU) General Product Safety Regulation (GPSR) is a set of rules that requires consumer products to be safe and our obligations to ensure this.

If you have any concerns about our products, you can contact us on ProductSafety@springernature.com

In case Publisher is established outside the EU, the EU authorized representative is:

Springer Nature Customer Service Center GmbH
Europaplatz 3
69115 Heidelberg, Germany

The manufacturer's authorised representative in the EU is Springer Nature Customer Service Centre GmbH, Europaplatz 3, 69115 Heidelberg, Germany. If you have any concerns regarding our products, please contact ProductSafety@springernature.com

Printed and bound by CPI Group (UK) Ltd, Croydon, CR0 4YY
01/10/2026
02225884-0003